ACC BOMBER TRIAD

By Don Logan

ROCKWELL B-1B: SAC's LAST BOMBER

NORTHROP's T-38 TALON

NORTHROP's YF-17 COBRA

THE 388TH TACTICAL FIGHTER WING AT KORAT RTAFB 1972

REPUBLIC's A-10 THUNDERBOLT II

BOEING C-135 SERIES

GENERAL DYNAMICS F-111 AARDVARK

ACC BOMBER TRIAD

(all titles are available from Schiffer Publishing Ltd.

ACC BOMBER TRIAD

The B-52s, B-1s, and B-2s of Air Combat Command

Don Logan

Schiffer Military History
Atglen, PA

ACKNOWLEDGMENTS

I would like to thank Brian C. Rogers and Tom Kaminski for their help in the research for this book. I also would like to thank Mark Shelly for his help proofing the manuscript and supplying suggestions for improvement of the text. Thanks to Roger Johansen for his work drawing the line art. The majority of the flight suit patches were supplied by John Cook and Rick Tolman.

I would also like to thank the following individuals who provided photographs included in this book: Don Abrahamson, David Brown, Tony Cassanova, George Cockle, Robert F. Dorr, Jerry Geer. Jim Geer, Norris Graser, Mike Grove, Paul Hart, R.E.F. Jones, Tom Kaminski, Craig Kaston, Ben Knowles, Nate Leong, Warren A. Montgomery, David Pasquini, Brian C. Rogers, Mick Roth, Douglas Slowiak/Vortex Photographics, Keith Snyder, Daniel Soulaine, Don Spering/AIR, Randy Walker, Jeff Wilson, Alan Winzerling, and Northrop Grumman.

THE AUTHOR

After graduating from California State University-Northridge with a BA degree in History, Don Logan joined the USAF in August 1969. He flew as an F-4E Weapon Systems Officer (WSO), stationed at Korat RTAFB in Thailand, flying 133 combat missions over North Vietnam, South Vietnam, and Laos before being shot down over North Vietnam on July 5, 1972. He spent nine months as a POW in Hanoi, North Vietnam. As a result of missions flown in Southeast Asia, he received The Distinguished Flying Cross, The Air Medal with twelve oak leaf clusters, and The Purple Heart. After his return to the U.S., he was assigned to Nellis AFB where he flew as a rightseater in the F-111A. He left the Air Force at the end of February 1977.

In March of 1977 Don went to work for North American Aircraft Division of Rockwell International, in Los Angeles, as a Flight Manual writer on the B-1A program. He was later made Editor of the Flight Manuals for B-1A #3 and B-1A #4. Following the cancellation of the B-1A production, he went to work for Northrop Aircraft as a fire control and ECM systems maintenance manual writer on the F-5 program.

In October of 1978 he started his employment at Boeing in Wichita, Kansas as a Flight Manual/ Weapon Delivery manual writer on the B-52 OAS/CMI (Offensive Avionics System/Cruise Missile Integration) program. He is presently the editor for Boeing's B-52 Flight and Weapon Delivery manuals.

Don Logan is also the author of *Rockwell B-1B: SAC's Last Bomber*, *The 388th Tactical Fighter Wing At Korat Royal Thai Air Force Base 1972*; *Northrop's T-38 Talon*; *Northrop's YF-17 Cobra*; *Republic's A-10 Thunderbolt II*; *The Boeing C-135 Series: Stratotanker, Stratolifter, and other Variants*; and *General Dynamics F-111 Aardvark*. (All available from Schiffer Publishing Ltd.)

Book Design by Robert Biondi.

Library of Congress Catalog Number: 98-88033.

Printed in China.
ISBN: 0-7643-0680-4

We are interested in hearing from authors with book ideas on military topics.

Published by Schiffer Publishing Ltd.
4880 Lower Valley Road
Atglen, PA 19310 USA
Phone: (610) 593-1777
FAX: (610) 593-2002
E-mail: Schifferbk@aol.com.
Visit our web site at: www.schifferbooks.com
Please write for a free catalog.
This book may be purchased from the publisher.
Please include $3.95 postage.
Try your bookstore first.

In Europe, Schiffer books are distributed by:
Bushwood Books
6 Marksbury Road
Kew Gardens
Surrey TW9 4JF
England
Phone: 44 (0)181 392-8585
FAX: 44 (0)181 392-9876
E-mail: Bushwd@aol.com.

Try your bookstore first.

Contents

(USAF)

Introduction

The United States Air Force is the only Air Force in the world with the ability to project power rapidly, precisely, and globally using strategic bombers. Originally all of today's USAF strategic bombers were part of the U.S. Nuclear Deterrent, planned to deliver nuclear weapons against targets in the Soviet Union. With the demise of the Soviet Union and the USAF reorganization, inactivating Strategic Air Command (SAC), the strategic bombers were moved to the new Air Combat Command (ACC). As part of ACC, USAF strategic bombers are continuing to evolve as a desirable option for response to a crisis anywhere in the world meeting the Global Attack needs of the USAF. Bombers provide the range, payload, lethality, and speed to guarantee the U.S. Air Force can globally project power anywhere, anytime.

With their global range, large payloads and immediate response capability, heavy bombers are at the core of the USAF's war fighting strategy. USAF bomber forces can employ while other forces deploy. Their ability to be the "first to the fight" allows bombers to destroy time-critical targets, attrit enemy forces rapidly, decisively halt advancing enemy forces, and prepare the battlefield for the counter attack. As the U.S. reduces the number of military forces permanently stationed overseas, its power projection capabilities will be even more important to the Joint Force Commander and its national leaders.

Today, the USAF bomber force, comprised of the B-52H, the B-1, and the B-2 provides a conventional deterrent option for the nation. Nuclear deterrence remains a basic requirement of national security, with the B-52Hs and B-2s providing the manned bomber portion of the U.S. nuclear weapon force.

The ACC Bomber Triad

Over the next several decades, with fewer overseas bases for the USAF, US national security will increasingly depend an conventional bombers to be able to strike targets in various regions of the world. No other resource in the USAF arsenal brings together the range, flexibility, and precise firepower as does the land-based bomber force. The United States shares a responsibility to defend its interests and those of its allies in many regions around the world. Long range bombers have strengths no other weapon system can match. Their combination of range, payload and flexibility make bombers the theater commander's weapon of choice both for response to a crisis and sustained combat operations. With advanced conventional munitions and other upgrades being added to the USAF bombers, they will continue as a viable strike force well into the next century.

Until recently, U.S. plans for its bomber force centered on nuclear deterrence against a strike by the Soviet Union, with conventional weapon delivery as secondary mission. With the breakup of the Soviet Union, the conventional mission became the primary role of the USAF Strategic Bomber force. With ever tighter U.S. Defense Budgets, it became necessary to set clear priorities for bomber upgrades. It was necessary to determine what modifications would be required for the B-1 and B-52, in part to compensate for decreasing the B-2 buy to 21 aircraft. Another task was to trim the B-52 force. Most of all it was necessary to increase conventional capabilities across the bomber fleet, and work on smart tactics to integrate the B-52s, B-1s, and B-2 into an effective bomber triad, managing the bomber force as a single entity.

The total USAF Bomber operational inventory of 208 aircraft is comprised of 94 B-52Hs, 93 B-1Bs, and 21 B-2s meets planned future requirements. With that force structure, bombers will be able to strike the priority target set under the harshest wartime conditions with a 35 hour mission, launching from and returning to continental United States bases. Not all of the total number of airframes will be available to theater commanders. Accounting for training requirements and backup inventory (depot maintenance, tests, etc.), the force structure provides theater commanders with a maximum of about 80 B-52Hs, 80 B-1Bs, and 16 B-2s. The B-52H and B-2 retain their nuclear potential as needed in accordance with the commander of STRATCOM's evaluation of future requirements. The B-1 will retain a limited nuclear capability, but will have no operational nuclear requirements, and therefore will not train for nuclear missions.

The bomber fleet is improving its long range navigation capabilities by adding GPS upgrades. The B-52 is improving its GPS system by adding a TACAN emulation software and displays for the pilots. The B-1 is adding GPS to its navigation and weapon delivery system.

The Joint Direct Attack Munition (JDAM) program grew out of Air Force and Navy requirements and provides direct attack capability based on an inertially-guided, Global Positioning System (GPS) assisted munition. The JDAM used in the bomber fleet (GBU-31(v)1/B and GBU-31(v)3/B) is a 2000 pound class weapon that will be accurate to within less than 10 feet of a target, day or night, in adverse weather. Even in regions like the Middle East, cloud cover can obstruct targets up to 30% of the time and for Northeast Asia the figure rises to as high as 60% in the worst weather seasons. JDAM with its GPS targeting has ability to overcome poor weather conditions which would render a laser or optically guided weapon ineffective.

The AGM-154 Joint Standoff Weapon (JSOW) is a glide weapon which provides accurate standoff capability, enabling bomber aircraft to launch outside the range of ground unit anti-aircraft defenses and achieve multiple kills per JSOW using sensor-fuzed submunitions.

The Tri-Service Standoff Attack Missile (TSSAM) originally planned for use in the USAF bomber force was canceled and replaced by the Joint Air to Surface Standoff Missile (JASSM). The JASSM, to be built by Lockheed-Martin, combines extended range (100 nm+) with an autonomous 1000 pound class precision warhead in a highly stealthy cruise

60-0050, assigned to the 419th Test Squadron at Edwards AFB is releasing a JDAM during a test mission. (USAF)

missile to yield tremendous flexibility and the ability to destroy small, high-value targets from outside the high-threat environment.

The CBU-103, CBU-104, and CBU-105 WCMD (Wind Corrected Munitions Dispenser) weapons can be carried by the B-52 and is being integrated in the B-1 and B-2. The WCMD has an inertially guided tail kit attached to the Tactical Munitions Dispenser (TMD) case, giving it greater accuracy than a gravity released unguided CBU. In addition, the guidance system allows the bombers to release WCMDs against a high threat targets without having to overfly the target.

The older, less glamorous weapons remain essential. The B-52, B-1 and B-2 carry conventional ordnance, such as the 500 pound Mk 82 and the 2000 pound Mk 84 bombs for use against a range of targets that do not demand precision delivery. All three bombers also have air-deliverable sea mine-laying capability to augment the Navy's sea denial capability, and the B-52Hs equipped with Harpoon missiles provide anti-ship capabilities.

B-52

The B-52 provides the National Command Authorities with the ability to respond to the full spectrum of conflict nuclear and conventional, at global distances. Integrated Conventional Stores Management System (ICSMS) was added to the B-52G and later to the B-52H. ICSMS tied the B-52's conventional weapon delivery system to the Offensive Avionics System (OAS) originally developed for delivery of nuclear weapons. The ICSMS modification greatly expanded the capabilities of the of the B-52 and made it the nation's conventional workhorse, giving it the ability to deliver the AGM-86C Conventional Air Launch Cruise Missile (CALCM), the precision optically guided AGM-142, and most importantly, the formidable ability to carry large numbers of most conventional gravity munitions in the U.S. inventory. The B-52Gs, modified with ICSMS, were credited with delivering over 30 percent of the total Desert Storm munitions tonnage, while comprising only 3 percent of the aircraft deployed.

The B-52 is gaining greater lethality and at the same time better survivability with precision gravity and stand-off weapons currently in development. The JDAM and WCMD were added to the B-52 in 1998, with the JSOW following in 2000, and JASSM is 2001. Additionally, a updated CALCM missile is in final testing. It will have an improved precision penetration capability, increasing the B-52's ability to destroy hardened targets.

B-1

The B-1, originally designed as a nuclear bomber, is undergoing the Conventional Mission Upgrade Program (CMUP) which transitions it to a conventional bomber. Addition of new conventional weapons and defensive system upgrades are improving the B-1's lethality and survivability, allowing it to respond on the first day of a conflict. Carrying 84 general purpose 500-pound bombs or 30 cluster bombs, the B-1 can help halt an enemy advance during the initial stages of a conflict while other forces deploy.

The B-1's conventional capability will continue to increase through the time-phased CMUP modernization effort. The advancements in B-1 weapon carriage significantly improve the platform's lethality against high, medium, and low threat targets. Programmed reliability and maintainability improvements, such as avionics computer replacement and defensive system upgrades presently underway, are designed to make the B-1 more affordable to operate. The Air Force is integrating advanced munitions into the B-1 fleet, with the Joint Direct Attack Munition (JDAM) being added in 1999, followed by the Wind Corrected Munitions Dispenser (WCMD), Joint Stand-off Weapon (JSOW), Joint Air-to-Surface Stand-off Missile (JASSM) by 2002.

B-2

The B-2's range and ability to penetrate defended airspace make a critical contribution to global power projection. The joint force commander can use the B-2 to strip away lethal

(USAF)

enemy air defenses, command and control nodes, and essential weapons of mass destruct on storage and production facilities.

Current B-2s incorporate the first generation avionics systems to include Terrain Following (TF) and Global Positioning System (GPS) navigation capability. Block 30 deliveries began in late Fiscal Year 1997 and provide low observable combat capabilities. The B-2's final stealth configuration will dramatically reduce its vulnerability to threat systems. Additionally, on-board systems such as TF, communications, situational awareness, and advanced targeting allow its pilots to attack the target while avoiding enemy defenses. By mid-2000, all B-2s will be Block 30 configured, making the entire B-2 fleet fully deployable. The Air Force is aggressively adding a formidable conventional mission capability to the B-2 allowing it to attack almost any target, anywhere, anytime. The fleet having been modified to the Block 20 configuration can currently employ the Mk-84, 2000-pound unguided conventional munition, the GBU-36/B GPS Aided Munition (GAM), up to 34 Cluster Bomb Units (CBUs), or 16, 2000 pound JDAMs GPS/inertially guided bombs. The GPS-Aided Targeting System (GATS) also contained in the Block 20 upgrade allows the B-2 partial low-level terrain following capability. The 509th Bomb Wing from Whiteman AFB, Missouri executed a GBU-36/B GAM operational demonstration on October 8, 1996. Sixteen GAMs were released from three B-2s destroying 16 different targets (8 ft x 8 ft x 20 ft steel boxes). The latest version of the B-2 (Block 30) caries JDAM (replacing GAM), three variants of Tactical Munitions Dispensers, Mk-62 Navy mines, 500-pound and 750-pound unguided bombs, new upgrades such as the GBU-37/B (GAM-113) hard target penetration munition, and JSOW. Planned post-Block 30 upgrades include JASSM.

BOMBER OPERATIONS IN OPERATION DESERT FOX

As part of the four day bombing campaign against Iraq whichstarted on December 16, 1998, U.S. Air Force B-52Hs and B-1Bsattacked part of the over 100 targets hit by U.S. and Britishforces.

The B-52H once again was used to launch AGM-86C ConventionalAir Launched Cruise Missiles (CALCM). Starting on the second day of the campaign, 15 B-52Hs flying out of Diego Garcia in the Indian Ocean launched approximately 90 CALCMs against targets in central and southern Iraq. Both Block 0 CALCMs (2,000 pound blast/fragmentation warhead) and Block 1 CALCMs (3,000 pound blast/fragmentation warhead) were used. This was the first time the more destructive Block 1 missiles were used. The following B-52Hs deployed to Diego Garcia in support of Operation Desert Fox:

60-0007	60-0010	60-0015	60-0016
60-0023	60-0025	60-0034	60-0036
60-0043	60-0044	60-0046	61-0006
61-0007	61-0020	61-0023	

The B-1Bs' introduction to combat also occurred on the second day of the campaign. Two B-1s flying from their base in Oman attacked Republican Guard barracks in west-northwest Iraq.

In a mission lasting approximately six hours, each bomber hit a different group of buildings in the complex. The B-1s flying at medium/high altitude each released 66 500 pound bombs against its target. A second pair of missions was flown on the third night of the campaign. The following B-1Bs deployed to Oman in support of OperationDesert Fox:

7th Bomb Wing Aircraft	28th Bomb Wing Aircraft
85-0067	85-0084
86-0109	86-0096
86-0135	86-0102

The following took part in the two mission sets flown in Operation Desert Fox

December 18, 1998 - First Mission
86-0096, 37th Bomb Squadron, 28th Bomb Wing, Ellsworth AFB, Callsign SLAM 01
86-0135, 9th Bomb Squadron, 7th Bomb Wing, Dyess AFB, Callsign SLAM 02

December 19, 1998 - Second Mission
86-0135, 9th Bomb Squadron, 7th Bomb Wing, Dyess AFB, Callsign SLAM 03
86-0102, 37th Bomb Squadron, 28th Bomb Wing, Ellsworth AFB, Callsign SLAM 04

B-52 Stratofortress

When ACC took over SAC's bomber assets in June, 1992, 84 B-52Gs and 95 B-52Hs were in service. The B-52 was to be used by ACC as a standoff weapon launcher and a precision bomber capable of delivering massive firepower in a low threat environment.

Boeing's B-52 Stratofortress which had been designed in the late 1940s and early 1950s, has served superbly for almost five decades as the mainstay of the United States' nuclear and conventional bomber force. The bomber is capable of flying at high subsonic speeds at altitudes up to 50,000 feet (15,166.6 meters). It can carry nuclear or conventional ordnance with worldwide precision navigation capability.

Starting in 1989, an on-going modification incorporated the global positioning system, heavy stores adapter beams for carrying heavy munitions and additional smart weapons capability for the B-52Gs. In addition B-52G aircraft were modified to carry the AGM-142 Have Nap missile and AGM-84 Harpoon anti-ship missile. Despite the B-52Gs' outstanding record, with the downsizing of the U.S. military, the decision was made to retain the slightly newer B-52Hs. The H model's TF-33 fanjet engines are 30% more powerful than the B-52G's engines which had to rely on water-injection for takeoff. In addition the TF-33s are more fuel efficient and cost less to operate, give the H model a greater range and, because they do not need the distilled water for water injection, can fight more easily from austere locations on short notice. Updated with modern technology, the B-52H will continue into the 21st century as an important element of our nation's defenses. Current engineering analysis show the B-52H's life span to extend beyond the year 2045.

The B-52Hs conversion to a conventional role brought advanced conventional capabilities into the bomber force. The B-52Hs gained new precision-standoff (conventional ALCM) and precision direct attack JDAM capabilities as part of the Conventional Enhancement Modification (CEM). These modifications included the addition of the MIL STD 1760 weapons data bus which enabled the aircraft to stand off and attack many targets early in a conflict, and operate in a more effective direct (precision) attack mode in later stages of the conflict.

On October 1, 1991, the B-52 Gunners were removed from the air crews. Advances in air to air weapons had made the gun system obsolete, and most opposing fighters carried missiles which could be launched outside the range of the B-52's tail gun. The guns and the fire control systems have been deactivated. All the guns had been removed from the B-52H's by mid 1995. The gunners station, next to the Electronic Warfare Officers station, remained and the ejection seat remained operational.

The retained B-52's fill two important roles in the ACC's arsenal. The first is as a standoff weapons carrier. War plans incorporate this capability early, to compensate for the increasing drawbacks of the B-52's large radar cross section and slower speed. Using conventional cruise missiles, as in Operation Desert Strike and Operation Desert Fox, the B-52Hs are able to strike fixed targets in defended areas at acceptable risk. In some cases, after U.S. forces deploy to a theater, the B-52H may be able to launch standoff weapons from airspace secured by joint air superiority assets. Second, just as with the B-52G in Desert Storm, the B-52H has the ability to deliver massive firepower in areas of lower threat. The B-52's ability to lift 38,000 pounds of bombs from bases anywhere in theater has grown to include precision weapons such as JDAM and WCMD. Ongoing upgrades such as GPS are enhancing both penetration and stand-off capabilities.

The Heavy Stores Adapter Beams (HSABs) on the under wing pylon stations allow the B-52H to carry heavy external weapons, a capability not shared by the B-1 or the B-2. The use of aerial refueling gives the B-52 a range limited only by crew endurance. It has an unrefueled combat range up to 8,800 miles (14,080 kilometers), depending on external weapon load. It is highly effective when used for ocean surveillance, and can assist the U.S. Navy in anti-ship and mine-laying operations. Two B-52s, in two hours, can monitor 140,000 square miles (364,000 square kilometers) of ocean surface.

All B-52Hs are equipped with an Electro-optical Viewing System (EVS) that uses platinum silicide forward-looking infrared and high resolution low-light-level television sensors to augment the targeting, battle assessment, flight safety, and terrain-avoidance system, thus further improving its combat ability and low-level flight capability. To enhance night capabilities, the two pilots can wear Night Vision Goggles (NVGs) giving improved night vision during low-level terrain-following operations. NVGs provide greater safety during night operations by increasing the pilot's ability to visually clear terrain and avoid enemy radar.

The B-52A first flew in 1954, and the B model entered service in 1955. A total of 744 B-52s were built with the last, B-52H 61-0040, delivered in October 1962. 193 B-52G models were delivered between October, 1958 and February, 1961. 102 B-52H aircraft were delivered between May, 1961 and October 1962. All of the B-52Gs and B-52Hs were built at Boeing's Wichita, Kansas facility. Only the H model is still in the Air Force inventory.

RAPID EIGHT

With the retiring of the B-52G's, the USAF lost its capability to launch AGM-84 and AGM-142 standoff weapons. A new program, Conventional Enhancement Modification (CEM), was started to modify B-52Hs and replace the lost B-52G capability. Aircraft modified by CEM would not be available until mid 1995. To fill the gap in lost capability a program called RAPID EIGHT was developed in early 1994, with a goal of having four B-52Hs modified for AGM-84 Harpoon carriage and launch and four B-52Hs modified for AGM-142 Have Nap carriage and launch as a "stop gap" measure until CEM aircraft were available. Modification of the four Rapid Harpoon aircraft (60-0013, 61-0013. 61-0019, and 61-0024) and the four Rapid Have Nap aircraft (60-0014, 60-0025, 60-0062, and 61-0004) was completed during the summer of 1994. As CEM modified aircraft became available the Rapid Eight aircraft were de-modified as they went through their depot (Tinker AFB) level maintenance.

CONVENTIONAL ENHANCEMENT MODIFICATION (CEM) PROGRAM

The Conventional Enhancement Modification program is a program planned to upgrade all B-52Hs with increased conventional weapon capability. Starting modification of aircraft during the summer of 1995. The CEM program added GPS and Military Standard 1760 (MIL-STD-1760) weapons data bus to the B-52H. In addition to modifications to the aircraft avionics system, stub pylons and Heavy Stores Adapter Beams were modified to allow carriage, programming, and release of MIL STD 1760 weapons. A new basic Offensive Avionics System (OAS) software was developed which allow special software packages called Stores Management Overlays (SMO) to be loaded. This allowed greater flexibility in weapon carriage. When a new weapon was to be added, a SMO would be built, without having to rewrite the existing software. SMOs have been written for AGM-84, AGM-86C, AGM-142, JDAM, and WCMD. In development are SMOs for JSOW and JASSM.

As of publication of this book fifty-three B-52Hs aircraft have been modified with CEM capability:

60-0003	60-0019	60-0041	60-0060	61-0016
60-0004	60-0020	60-0043	60-0062	61-0020
60-0007	60-0022	60-0043	61-0002	61-0021
60-0009	60-0023	60-0046	61-0003	61-0022
60-0010	60-0025	60-0049	61-0004	61-0023
60-0011	60-0026	60-0050	61-0005	61-0027
60-0014	60-0031	60-0051	61-0007	61-0029
60-0015	60-0032	60-0052	61-0008	61-0031
60-0016	60-0033	60-0057	61-0010	61-0038
60-0017	60-0034	60-0058	61-0011	61-0039
60-0018	60-0037	60-0059		

B-52 General Characteristics

Engines: Eight Pratt & Whitney engines TF33-P-3/103 turbofan
Thrust: Each engine up to 17,000 pounds
Length: 159 feet, 4 inches (48.5 meters)
Height: 40 feet, 8 inches (12.4 meters)
Wingspan: 185 feet (56.4 meters)
Speed: 650 miles per hour (Mach 0.86)
Ceiling: 50,000 feet (15,151.5 meters)
Weight: Approximately 185,000 pounds empty (83,250 kilograms)
Maximum Takeoff Weight: 488,000 pounds (219,600 kilograms)
Range: Unrefueled 8,800 miles (7,652 nautical miles)

Armament: Approximately 70,000 pounds (31,500 kilograms) mixed ordnance – bombs, mines and missiles. (Modified to carry air-launched cruise missiles, Harpoon anti-ship and Have Nap missiles.)
Crew: Five (aircraft commander(pilot),copilot, radar navigator, navigator and electronic warfare officer)
Accommodations: Six ejection seats (four ejecting upward and two ejecting downward)
Present Inventory:
Active force, 85 (at two bases -Barksdale AFB, Louisiana and Minot AFB, North Dakota)
Air National Guard, 0
Air Force Reserve Command, 9 (all assigned to Barksdale AFB, Louisiana)

B-52 OPERATIONAL UNITS

The B-52's were assigned to nine SAC Bomb Wings when transferred to the new Air Combat Command in June 1992. The B-52Gs were assigned to the 2nd Bomb Wing at Barksdale AFB, Louisiana; the 42nd Bomb Wing at Loring AFB, Maine; 93rd Bomb Wing at Castle AFB, California; the 379th Bomb Wing at Wurtsmith AFB, Michigan; and nine B-52Gs were assigned to the 34th Bomb Squadron, 366th Operations Group at Castle AFB, California. The 34th Bomb Squadron was a Geographically Separated Unit (GSU) of the 366th Wing Based at Mountain Home AFB, Idaho. The B-52Hs were assigned to the 5th Bomb Wing at Minot AFB, North Dakota; the 7th Bomb Wing at Carswell AFB Texas; the 92nd Bomb Wing, Fairchild AFB, Washington; the 410th Bomb Wing at K.I. Sawyer AFB, Michigan; and the 416th Bomb Wing at Griffiss AFB, New York.

As the B-52Gs were retired, and additional bomber bases were closed, the remaining B-52Hs were consolidated at two bases, Barksdale AFB, Louisiana, and Minot AFB, North Dakota. At Barksdale AFB, the 2nd Bomb Wing replaced their B-52Gs with B-52Hs, and the 917th Wing of Air Force Reserve Command became the first and only Air Force Reserve unit to operate the B-52. The 5th Bomb Wing at Minot AFB, North Dakota kept their B-52Hs. Presently all operational B-52s are assigned to these two bases.

B-52Hs on the flight line at Griffiss AFB on April 1, 1994. (Tom Kaminski)

BARKSDALE AFB, LOUISIANA

8th AIR FORCE HEADQUARTERS

The 8th Air Force headquartered at Barksdale AFB is the Numbered Air Force responsible for the operation of the long range bombers in the active USAF inventory. Assigned to the 8th AF are: the 2nd Bomb Wing, based at Barksdale AFB and operating B-52Hs; the 5th Bomb Wing based at Minot AFB, North Dakota and operating B-52Hs; the 7th Bomb Wing, based at Dyess AFB, Texas operating B-1s, the 27th Fighter Wing, based at Cannon AFB operating F-16Cs and Ds (the 27th Fighter Wing had been ACC's F-111 wing until the last 111s were retired during 1996); the 28th Bomb Wing, based at Ellsworth AFB, South Dakota operating B-1s; and the 509th Bomb Wing, based at Whiteman AFB, Missouri, the only wing operating the B-2. Additionally two support units, the 65th Air Base Wing at Lajes Field, the Azores; and the 85th Group at NAS Keflavik, Iceland are assigned to the 8th Air Force. The only "Active Duty" bomber unit not assigned to the 8th Air Force is the 34th Bomb Squadron. It is a B-1 squadron assigned to the 366th Wing at Mountain Home AFB, Idaho, and part of the 12th Air Force headquartered at Davis Monthan AFB, Arizona.

The tail of B-52H 60-0008 seen on September 26, 1993, with the red tail cap of the 96th Bomb Squadron and marked as the 8th Air Force Flagship. (Author)

2nd BOMB WING

The 2nd Bomb Wing moved to Barksdale AFB, Louisiana on April 1, 1963 from Hunter AFB, Georgia. It took over the equipment (B-52F) and personnel of the 4238th Strategic Wing (SW). It currently operates the B-52H, with three squadrons; the 11th Bomb Squadron, the 20th Bomb Squadron, and the 96th Bomb Squadron. In addition the 2nd Bomb Wing supports the B-52 Weapons School (Detachment 5, USAF Weapons School, 57th Wing) and the 49th Test and Evaluation Squadron (assigned to the 53rd Wing at Eglin AFB), both located at Barksdale AFB. The 2nd Bomb Wing's B-52Hs carry a LA (**L**ouisian**A**) tail code, with a different color tail stripe for each squadron. Within the tail stripe for all squadrons is a Fleur-de-lis and the word "BARKSDALE" in black. The Wing's motto is Libertatem Defendimus (We Defend Liberty).

The tail of B-52H 60-0001 seen on September 26, 1993, with the blue tail cap of the 20th Bomb Squadron and marked as the 2nd Bomb Wing Flagship. (Author)

11th Bomb Squadron

The 11th Bomb Squadron (BS) was activated on July 1, 1994 as part of the consolidation which moved all operational B-52Hs to two bases (Barksdale AFB and Minot AFB). The 11th BS is also tasked with the FTU (Formal Training Unit) duties; training B-52 air crews. It is currently active flying B-52Hs as part of the 2nd Bomb Wing. The 11th Bomb Squadron's aircraft wear a gold tail stripe.

The tail of B-52H 61-0011 seen on August 5, 1993, with the yellow tail cap of the 11th Bomb Squadron and marked as the 11th Bomb Squadron Flagship. (Brian C. Rogers)

20th Bomb Squadron

The 20th Bomb Squadron operated as part of the 2nd Bomb Wing at Barksdale from April 1, 1963 until reassigned to the 7th Bomb Wing at Carswell AFB, Texas on June 25, 1965. The 20th BS was reassigned back to the 2nd Bomb Wing, moving back to Barksdale AFB on December 18, 1992, replacing the 62nd Bomb Squadron. Aircraft of the 20th Bomb Squadron are equipped with the AGM-142 Have Nap Missile. The 20th BS is currently active flying B-52Hs as part of the 2nd Bomb Wing. The 20th Bomb Squadron's aircraft wear a blue tail stripe.

The tail of B-52H 60-0020 seen on September 26, 1993, with the blue tail cap of the 20th Bomb Squadron and marked as the 20th Bomb Squadron Flagship. (Author)

62nd Bomb Squadron

The 62nd Bomb Squadron flew B-52Gs as part of the 2nd Bomb Wing from June 25, 1965 until replaced by the 20th Bomb Squadron on December 18, 1992. The 62nd was inactivated on January 18, 1993. The 62nd Bomb Squadron's aircraft wore a blue tail stripe.

The tail of B-52H 61-0002 seen on May 5, 1995, with the blue tail cap of the 20th Bomb Squadron and marked as the 2nd Operations Group (OG) Flagship. (Author)

96th Bomb Squadron

The 96th Bomb Squadron was activated on October 1, 1993, and replaced the 596th Bomb Squadron. The 96th is assigned both the AGM-84 Harpoon and the Conventional ALCM missions. Aircraft and crews of the 96th flew the CALCM strike missions against Iraq on September 3, 1996, the first time the B-52H launched weapons in actual combat. The 96th BS is currently active flying B-52Hs as part of the 2nd Bomb Wing. The 96th Bomb Squadron's aircraft wear a red tail stripe.

The tail of B-52H 60-0008 seen on May 2, 1994, with the red tail cap of the 96th Bomb Squadron and marked as the 96th Bomb Squadron Flagship. (Author)

596th Bomb Squadron

The 596th Bomb Squadron flew B-52Gs as part of the 2nd Bomb Wing from April 15, 1968 until the units G models were replaced with B-52Hs in 1992. The 596th BS was replaced by the 96th Bomb Squadron on October 1, 1992.

57th WING, DETACHMENT 5, B-52 DIVISION

In June 1992, following the inactivation of SAC, the B-52 Weapons School at Ellsworth AFB was inactivated and replaced by Detachment 1 of the USAF Weapons School, which included both B-52 and B-1 divisions. The B-52 Division moved to Barksdale AFB during 1995. As the 99th Wing was inactivating the two bomber divisions picked up the test functions previously accomplished by the 99th Wing. As a result both Divisions, having both Weapons School and Test functions, became part of the 57th Wing at Nellis AFB. The B-52 Division at Barksdale became Detachment 5, 57th Wing. The B-52 Weapons School does not have any aircraft of their own and borrow 2nd Bomb Wing aircraft when necessary.

57th Wing patch 1997 to present

57th Wing patch June 1993 to 1997

Current patch for the B-52 Weapon School at Barksdale AFB, used since November 1997

Used at Barksdale AFB 1995 to October 1, 1996

Used when the B-52 Weapons School was based at Ellsworth AFB (June 1, 1992) and Barksdale until replaced in November 1997

53rd WING, 79th TEST AND EVALUATION GROUP, 49th TEST SQUADRON

B-52 operational test and evaluation is accomplished by the 49th Test and Evaluation Squadron at Barksdale, part of the 79th Test and Evaluation Group (TEG), 53rd Wing which is based at Eglin AFB, Florida. The transfer of test functions occurred when the Air Warfare Center restructure took place in October 1996. As part of the restructure, the test functions for B-52 were transferred from the 57th Wing, Detachment 5 to the 49th TEST Squadron. The 49th TEST squadron does not have any aircraft of its own and borrows aircraft from other units when necessary.

917th WING AIR FORCE RESERVE COMMAND

Assigned to the 10th Air Force headquartered at NAS Fort Worth Joint Reserve Base, Carswell Field, Texas, the 917th Wing operates the Reserves' only long range bombers. On October 1, 1993, the 917th Fighter Wing was redesignated as the 917th Wing. On that date, the 46th Fighter Squadron, one on the unit's two A-10 squadrons, was inactivated and replaced by the 93rd Bomb Squadron. The 93rd BS B-52Hs were the first Strategic Bombers ever assigned to an Air Force Reserve unit. The 917th continues to operate B-52Hs at Barksdale AFB. The 917th Wing's B-52Hs carry a BD (**B**arks**D**ale) tail code, a blue and gold checked tail stripe, and an Indian head (from the 93rd Bomb Squadron insignia) on the wing external fuel tanks.

93rd Bomb Squadron

The 93rd was activated on October 1, 1993, replacing the 46th Fighter Training Squadron as the 917th's second flying squadron. The 93rd began receiving its B-52Hs on December 12, 1993, and is currently assigned nine B-52Hs.

The external fuel tanks of the 93rd Bomb Squadrons B-52Hs carry the Indian head from the 93rd patch as seen here on December 10, 1994. (Randy Walker)

ABOVE AND RIGHT: On April 1, 1997 the U.S. Air Force Reserve (AFRES) was reorganized as the Air Force Reserve Command (AFRC), changing the tail markings as seen here on B-52H 60-0003. (Author)

CARSWELL AFB, TEXAS

B-52H 60-0007, the only Carswell 7th Bomb Wing B-52H to wear the CW tail codes, is seen here on November 30, 1992. (Randy Walker)

7th Bombardment Wing

The 7th Bomb Wing converted from the B-36 to the B-52 during 1958. It operated B-52s from Carswell AFB, Fort Worth, Texas until December 18, 1992, when the last B-52H moved to Barksdale. The 7th Bomb Wing was inactivated on January 1, 1993. The Wing moved to Dyess AFB, Texas without personnel or equipment on October 1, 1993, to replace the 96th Bomb Wing. The 7th Bomb Wing presently operates B-1s at Dyess. The 7th Bomb Wing B-52H's wore a blue, green, and gray tail stripe containing a black steer head and FORT WORTH TEXAS in light gray. Just before the aircraft was transferred to the 2nd Bomb Wing, CW (**C**ars**W**ell) tail codes were added to 60-0007, along with a blue tail stripe containing FORT WORTH. The Wing's motto is Mors Ab Alto (Death From Above).

9th Bomb Squadron

The 9th Bomb Squadron was equipped with the B-52F from June, 1958 until inactivated on June 25, 1968. It was activated on December 31, 1971, assigned B-52Ds which it operated until they were replaced with B-52Hs in 1982. The squadron's aircraft transferred to the 2nd Bomb Wing at Barksdale during 1992, and the squadron was inactivated on August 15, 1992. The 9th Bomb Squadron was activated at Dyess AFB on October 1, 1993 replacing the 337th Bomb Squadron as a B-1 Squadron.

(Randy Walker)

20th Bomb Squadron

The 20th Bomb Squadron was equipped with the B-52F when it transferred from the 2nd BW at Barksdale on June 25, 1965. The 20th BS was assigned B-52Ds in 1969, which it operated until they were replaced with B-52Hs in 1983. The squadron's aircraft transferred to the 2nd Bomb Wing at Barksdale during 1992. The 20th Bomb Squadron moved to Barksdale AFB on December 18, 1992 replacing the 62nd Bomb Squadron, and operating as a B-52H Squadron.

CASTLE AFB, CALIFORNIA

93rd BOMB WING

The 93rd Bomb Wing received its first B-52s beginning in June 1955 and continued to operate B-52s until it retired its last B-52G in May 1994. The 93rd Bomb Wing at Castle was SAC's primary air crew training unit, training crews for both the B-52 and the KC-135. The 93rd BW ceased flight operations in May 3, 1994, and supervised the closing of Castle AFB during 1995. The 93rd Bomb Wing inactivated on September 30, 1995 and activated as the 93rd Air Control Wing (ACW) on January 29, 1996, at Robins AFB, Georgia. The 93rd ACW presently operates the E-8 Joint STARS aircraft. The 93rd Bomb Wing B-52Gs carried CA (**CA**stle) tail codes and blue tail stripes.

328th Bomb Squadron

The 328th Bomb Squadron operated B-52s at Castle AFB from June 1955 until it was inactivated on June 15, 1994. The 328th BS operated B-52Bs from June 1956 until 1965; B-52Ds from June 1956 to 1958, and again during 1965 through 1971; B-52Es during 1957 and 1958 also 1967 through 1970; B-52Fs 1958 through 1974, B-52Gs during 1967 and 1968, then again in 1974 through 1994; and the B-52H from 1974 through 1983. With the upcoming closure of Castle AFB and the retirement of the B-52G, the 328th Bomb Squadron was inactivated on June 15, 1994.

The tail of B-52G 59-2588 seen on November 8, 1992, with the blue tail cap of the 328th Bomb Squadron and marked as the 328th Bomb Squadron Flagship. (Craig Kaston)

329th Combat Crew Training Squadron

The 329th Combat Crew Training Squadron (CCTS) was assigned to the 93rd Bomb Wing at Castle AFB. It was activated as the 329th Strategic Bombardment Training Squadron (SBTS) on July 1, 1986. It was redesignated the 329th CCTS on January 1, 1987. With the upcoming closure of Castle AFB and the retirement of the B-52G, the 323th Bomb Squadron was inactivated on July 17, 1994.

34th Bomb Squadron

In June 1992, as part of Air Force restructuring, Air Combat Command replaced Tactical Air Command. A month later, on July 1, 1992, the 366th also gained the 34th Bomb Squadron. The 34th Bomb Squadron, though assigned to the 366th Wing at Mountain Home AFB, Idaho, was located at Castle AFB, California. The 34th BS flew the B-52G Stratofortress, giving the 366th Wing deep interdiction bombing capabilities, and was the only B-52 unit armed with the deadly, medium-range Have Nap missile. On April 4, 1994, with the retiring of the B-52Gs and the closure of Castle AFB, the 34th Bomb Squadron transferred its flag to Ellsworth AFB, South Dakota, making way for the squadron to be equipped with the B-1 Lancer. The B-52Gs of the 34th Bomb Squadron carried MO (**MO**untian Home) tail codes, and a tail stripe of a red thunderbird followed by red lines over a black stripe.

The tail of B-52G 57-6520 seen in August 1993 with the red and black tail cap of the 34th Bomb Squadron and marked as the 34th Bomb Squadron Flagship. (Author)

FAIRCHILD AFB, WASHINGTON

92nd BOMB WING

The 92nd Bomb Wing received its first B-52s during 1957, converting from the B-36. The 92nd Bomb Wing continued to operate B-52s until it transferred its last B-52H left Fairchild in mid 1994. The 92nd Bomb Wing was redesignated as the 92nd Air Refueling Wing on July 1, 1994. The B-52Hs carried a blue and green sea hawk on their tails. This was replaced by FC (**F**air**C**hild) tail codes and a blue tail stripe. During early 1993 the blue tail stripe was replaced by a red and black checked tail stripe. Despite the Wing's motto, Duplum Incolumitatis (Double Safety), the 92nd Bomb Wing was the only ACC unit to crash a B-52.

The tail of B-52H 61-0036 seen on May 2, 1994, with the red and black checkered tail cap of the 325th Bomb Squadron, 92nd Bomb Wing. (Author)

325th Bomb Squadron

The 325th Bomb Squadron operated B-52Ds from March 1957 until 1971. B-52Cs were also assigned during 1967 through 1971, with the B-52D being the Wing's primary aircraft. It began receiving B-52Gs in 1970. The squadron's B-52Hs arrived in 1985, remaining until the squadron was inactivated on July 1, 1994.

GRIFFISS AFB, NEW YORK

416th BOMB WING

The 416th Bomb Wing was activated at Griffiss AFB, New York on February 1, 1963, taking over aircraft and personnel of the 4039th Strategic Wing (SW). The Wing transferred its aircraft to other bases during the second half of 1994. The 416th Operations Group was inactivated on January 1, 1995, followed by inactivation of the 416th Bomb Wing on September 30, 1995, as a result of the closure of Griffiss AFB. The B-52Hs of the 416th Bomb Wing wore a GR (**GR**iffiss) tail code and a red tail stripe containing the words GRIFFISS in white.

668th Bomb Squadron

The 668th Bomb Squadron received B-52Gs from the 75th BS of the 4039th SW on February 2, 1963. The 668th converted to the B-52H in 1991. The squadrons last two B-52Hs were transferred to the 5th BW at Minot AFB on November 15, 1994. The squadron was inactivated on January 1, 1995.

The tail of B-52H 60-0030 seen on October 18, 1993, with the red and white tail cap of the 416th Bomb Wing and marked as the 668th Bomb Squadron Flagship. (Brian C. Rogers)

K.I. SAWYER AFB, MICHIGAN

The tail of B-52H 60-0011 seen on July 16, 1994, with the rainbow tail cap and eagle of the 410th Bomb Wing. (Author)

410th BOMB WING

The 410th Bomb Wing arrived at K.I. Sawyer AFB, Michigan on February 1, 1963, taking over the B-52H aircraft and personnel of the 4042nd Strategic Wing (SW). It continued to operate the B-52Hs until transferring them to the 2nd Bomb Wing and 5th Bomb Wing during the second half of 1994. The 410th Operations Group was inactivated on November 21, 1994, followed by inactivation of the 410th Bomb Wing on September 30, 1995, as a result of the closure of K.I. Sawyer AFB. The B-52Hs of the 410th BW carried KI (**K.I.** Sawyer) tail codes with a rainbow tail stripe containing a black eagle.

644th Bomb Squadron

The 644th Bomb Squadron arrived at K.I. Sawyer on February 1, 1963, acquiring the B-52Hs of the 526th Bomb Squadron, 4042nd SW. It operated B-52Hs until its inactivation on November 21, 1994. The last two B-52Hs were transferred to the 5th BW on November 15, 1994.

LORING AFB, MAINE

42nd BOMB WING

The 42nd Bomb Wing converted from B-36s to B-52s during 1956. The Wing continued to operate B-52s until it retired its B-52Gs during 1993 through 1994. The 42nd BW B-52Gs wore a LZ (**KLIZ** was the ICAO designator for Loring AFB) tail code, and a black tail stripe edged in red with "42d BOMB WING" in yellow. The Wing was inactivated on September 30, 1994. The Wing's motto is Aethera Nobis (With Us The Heavens).

69th Bomb Squadron

The 69th Bomb Squadron was equipped with B-52Cs in June 1956, replaced them with B-52Ds in 1957, and finally with B-52Gs in 1959. The 69th BS operated B-52Gs through 1994. The squadron was inactivated on December 31, 1993. The last B-52G departed for AMARC and retirement on March 8, 1994.

The tail of B-52G 58-0253 seen on June 21, 1993, with the black tail cap of the 42nd Bomb Wing. (Author)

42nd Bomb Wing

69th Bomb Squadron

MINOT AFB, NORTH DAKOTA

5th BOMB WING

In July 1961, the first B-52H Stratofortress, named PEACE PERSUADER, arrived on base and within five months the base received its first Hound Dog Missile to give the B-52 its first stand-off missile capability. Minot AFB's organizational makeup changed on July 25, 1968, when the 5th Bombardment Wing (BMW) replaced the 450th Bombardment Wing. This change was in line with Air Force policy in keeping active those units with the most illustrious histories. The 5 BMW continued to modernize in October 1989, when they added the Air Launched Cruise Missile (ALCM) to their B-52s. The 5th Bomb Wing gained AGM-129 Advanced Cruise Missiles (ACM) for their B-52s in the spring of 1993. The Wing's motto is Kiai O Ka Lewa (Guardians Of The Upper Realm)

The 5th Bomb Wing carries MT (**M**ino**T**) tail codes. The original 5th Bomb wing tail stripe was a red and white candy cane stripe edged in blue. This stripe was later changed to Squadron specific tail stripes.

The tail of B-52H 60-0005 seen on April 12, 1995, with the red and white candy cane stripe tail cap of the 5th Bomb Wing and marked as the 5th Bomb Wing Flagship. (Brian C. Rogers)

The tail of B-52H 60-0055 seen on March 30, 1996, with the red and yellow tail cap of the 23rd Bomb Squadron and marked as the 5th Bomb Wing Flagship. (Brian C. Rogers)

The tail of B-52H 61-0005 seen on June 23, 1995, with the black tail cap of the 72nd Bomb Squadron and marked as the 5th Operations Group (OG) Flagship. (Brian C. Rogers)

The tail of B-52H 60-0055 seen on August 27, 1996, with the red and yellow tail cap of the 23rd Bomb Squadron and marked as the 5th Operations Group (OG) Flagship. (Brian C. Rogers)

23rd Bomb Squadron

The 23rd Bomb Squadron moved to Minot as part of the 5th Bomb Wing transfer on July 25, 1968, taking over the B-52Hs and other assets of the 720th Bomb Squadron. The unit continues to fly B-52Hs as sole bomb squadron of the 5th Bomb Wing. The tail stripe is a red stripe edged in yellow with two bombs followed by the name "BARONS" and then three bombs, all in yellow. The two bombs and three bombs signify the 23rd Bomb Squadron, and the total of five signifys the 5th Bomb Wing. A tail stripe with the red and yellow colors reversed was tested on 61-0040, but was later removed.

72nd Bomb Squadron

The 72nd Bomb Squadron was activated on December 1, 1994, as a part of the realignment of B-52Hs which concentrated them at only Minot AFB and Barksdale AFB. It was assigned to the 5th Bomb Wing, and was operational from January 6, 1995 until December 31, 1995. The 72nd flew its last mission on December 21, 1995, and was inactivated on July 1, 1996. The Squadron tail stripe was a black stripe containing two white lightning bolts in the shape of "72."

The tail of B-52H 60-0023 seen on March 30, 1996, with the red and yellow tail cap of the 23rd Bomb Squadron and marked as the 23rd Bomb Squadron Flagship. (Brian C. Rogers)

The tail of B-52H 60-0061 seen on August 10, 1995, with the black tail cap of the 72nd Bomb Squadron and marked as the 72nd Bomb Squadron Flagship. (Brian C. Rogers)

WURTSMITH AFB, MICHIGAN

379th BOMB WING

The 379th Bomb Wing moved from Homestead AFB, Florida on January 9, 1961 to Wurtsmith AFB, Michigan. During 1961 the Wing received factory fresh new B-52Hs. The H models were traded for B-52Gs in 1977. The Wing ceased flying its B-52Gs during 1992, with the 379th Operations Group inactivating on December 31, 1992. The 379th Wing was inactivated on June 15, 1993. The 379th's B-52Gs did not carry tail codes, but rather a black letter "K" in a gray triangle edged with black. This Triangle K was the tail marking carried by the 379th Bomb Group's bombers during World War II. The Wing's motto is Diligentia Et Accuratio (Diligent and Accurate).

The tail of B-52G 58-0173 seen with the black and gray triangle (delta) K tail markings of the 379th Bomb Wing. (Author)

379th Bomb Wing

524th Bomb Squadron

The 524th Bomb Squadron was equipped with B-52Hs from May 1961 until mid 1977, when it converted to the B-52G. It continued to operate B-52Gs until the aircraft were retired. The Squadron was inactivated on December 15, 1992.

524th Bomb Squadron

B-52G

The G model of the B-52 was assigned to ACC from June 1, 1992 until the last G (58-0240) was sent into retirement at Aerospace Maintenance and Regeneration Center (AMARC) Davis-Monthan AFB on May 3, 1994. Most of the 100 cruise missile capable (CMI) B-52Gs had already been retired when ACC took over SAC's bomber force. Only the 2nd Bomb Wing at Barksdale AFB, Louisiana, and the 379th Bomb Wing at Wurtsmith AFB, Michigan were still operating CMI B-52Gs in June 1992. The majority of ACC's G models were Integrated Conventional Stores management System (ICSMS) equipped.

In addition to the ICSMS equipment, the all conventional G models had the capability to launch AGM-84 Harpoon Anti-Shipping missiles. Eight Gs, all assigned to the 34th Bomb Squadron, were also equipped with AGM-142 optically guided standoff missiles.

The G models continued in service until finally retired in the first half of 1994. Of ACC's B-52G units only the 2nd Bomb Wing now flying B-52Hs from Barksdale AFB and the 34th Bomb Squadron now flying B-1s from Mountain Home AFB are still active.

Because of the short, less than two years, assignment of B-52Gs to ACC units, photographs of all B-52Gs are not available. As a result, instead of providing a photographic coverage of each ACC B-52G as done in this book with ACC's other bombers, only examples showing unit markings are included. ACC's B-52Gs are listed by the unit assigned including serial number, weapon configuration, and date retired.

The last B-52G retired, 58-0240 is seen here on May 3, 1994, taxiing out for takeoff at Castle AFB on its retirement flight to AMARC. (W.A. Montgomery)

B-52G STRATOFORTRESS ACC ASSIGNMENTS

WEAPON CONFIGURATION CODES

H = AGM-84 Harpoon Capable
I = Integrated Conventional Stores Management System (ICSMS)
N = AGM-142 Have Nap Capable
CMI = Cruise Missile Integration (AGM-86 Cruise Missile Capable)

2nd Bomb Wing, Barksdale AFB, Louisiana

Serial Number	Weapon Configuration	Date Retired
57-6471	CMI	July 29, 1992
57-6472	CMI	July 8, 1992
57-6480	CMI	August 5, 1992
57-6490	CMI	December 1, 1992
57-6495	CMI	August 26, 1992
57-6498	CMI	November 3, 1992
57-6503	CMI	August 19, 1992
57-6508	CMI	July 22, 1992
57-6511	CMI	October 6, 1992
57-6515	CMI	September 30, 1992
58-0160	CMI	November 19, 1992
58-0166	CMI	December 21, 1992
58-0170	CMI	November 10, 1992
58-0173	CMI	August 5, 1992
58-0179	CMI	December 10, 1992
58-0181	CMI	September 23, 1992
58-0193	CMI	December 8, 1992
58-0211	CMI	September 2, 1992
58-0222	CMI	August 12, 1992
58-0227	CMI	July 15, 1992
58-0229	CMI	November 12, 1992
58-0236	CMI	October 13, 1992
58-0239	CMI	September 9, 1992
58-0244	CMI	October 29, 1992
58-0245	CMI	October 20, 1992
59-2566	CMI	September 16, 1992
59-2591	CMI	June 19, 1992
59-2594	CMI	October 15, 1992
59-2602	CMI	October 27, 1992

B-52G 58-0166 is seen here at AMARC on March 10, 1993, shortly after arrival. The rounded leading edge at the Wing root indicates this was CMI equipped. (Norris Graser)

B-52G 658-0179 is seen here at AMARC on March 13, 1993, shortly after arrival. This aircraft is also CMI equipped. (Norris Graser)

42nd Bomb Wing, Loring AFB, Maine

Serial Number	Weapon Configuration	Date Retired
57-6488	H I	July 29, 1993
58-0195	H I	November 18, 1993
58-0197	H I	November 16, 1993
58-0221	H I	October 21, 1993
58-0226	H I	October 19, 1993
58-0230	H I	May 6, 1993
58-0250	H I	October 7, 1993
58-0253	H I	December 4, 1993
58-0255	H I	October 28, 1993
58-0257	H I	October 14, 1993
59-2573	H I	October 26, 1993
59-2585	H I	April 15, 1993

B-52G 58-0197 is seen here at Loring AFB on August 9, 1993. (Ben Knowles)

B-52G 58-0226 is seen here at Loring AFB on August 9, 1993. (Daniel Soulaine)

B-52G 58-0250 is seen here at Loring AFB on August 9, 1993. (Ben Knowles)

B-52G 58-0253 is seen here at Ellsworth AFB on June 21, 1993, as one of the B-52s deployed to support flying activities of the Bomber Weapons School. (Author)

B-52G 58-0197 is seen here at Loring AFB on August 9, 1993. (Daniel Soulaine)

93rd Bomb Wing, Castle AFB, California

Serial Number	Weapon Configuration	Date Retired
57-6473	H I	February 25, 1993
57-6476	H I	November 2, 1993
58-0163	H I	April 14, 1994
58-0192	H I	February 10, 1994
58-0206	H I	February 1, 1994
58-0210	H I	March 30, 1993
58-0213	H I	November 23, 1993
58-0214	H I	April 12, 1994
58-0216	H I	March 8, 1994
58-0218	H I	January 18, 1994
58-0233	H I	February 17, 1994
58-0240	H I	May 3, 1994
58-0248	H I	January 25, 1994
58-0258	H I	November 9, 1993
59-2565	H I	October 12, 1993
59-2572	H I	January 11, 1994
59-2588	H I	April 5, 1994
59-2595	H I	February 24, 1994
59-2599	H I	March 15, 1994

B-52G 58-0163 is seen here marked as the 93rd Operations Group (OG) Flagship. (Ben Knowles)

B-52G 58-0214 is seen here at Nellis AFB during October 1993. (David Brown)

B-52G 58-0240 is seen here at Nellis AFB on October 14, 1993, marked as the 93rd Bomb Wing Flagship. (Brian C. Rogers)

B-52G 58-0248, in dark gray paint, is seen here at Nellis AFB on October 14, 1993. The landing drag chute is visible, still attached to the tail. (Brian C. Rogers)

B-52G 59-2565 is seen here at AMARC on November 27, 1993, a little more than a month after it was retired. (Douglas Slowiak)

B-52G 59-2588 is seen here at Castle AFB on May 27, 1993, marked as the 328th Bomb Squadron Flagship. (Author)

B-52G 59-2599, in dark gray paint, is seen here at Castle AFB on May 27, 1993. (Brian C. Rogers)

B-52G 59-2588 is seen here on landing with its drag chute deployed, and marked as the 328th Bomb Squadron Flagship. (Ben Knowles)

366th Wing, 34th Bomb Squadron

Serial Number	Weapon Configuration	Date Retired
57-6497	H I	January 20, 1994
57-6520	HNI	January 27, 1994
58-0202	HNI	March 10, 1994
58-0203	HNI	March 3, 1994
58-0212	HNI	February 8, 1994
58-0242	HNI	February 22, 1994
59-2569	HNI	February 15, 1994
59-2570	HNI	January 24, 1994
59-2598	HNI	January 13, 1994

B-52G 57-6520 is seen here at Castle AFB on May 27, 1993, marked as the 34th Bomb Squadron Flagship. (Author)

B-52G 58-0203 is seen here in 34th Bomb Squadron markings on landing approach at Castle AFB during August 1993. (Ben Knowles)

RIGHT: B-52G 58-0211 is seen here in 34th Bomb Squadron markings at Castle AFB on May 27, 1993. (Author)

B-52G 59-2569 is seen here in 34th Bomb Squadron markings with an AGM-142 Have Nap missile of the forward station of the left Wing pylon. (Ben Knowles)

B-52G 59-2570 is seen here in 34th Bomb Squadron markings at Castle AFB on May 27, 1993. (Author)

B-52G 59-2598 is seen here in 34th Bomb Squadron markings at Castle AFB on May 27, 1993. (Author)

379th Bomb Wing, Wurtsmith AFB, Michigan

Serial Number	Weapon Configuration	Date Retired
57-6492	CMI	December 15, 1992
58-0164	CMI	November 5, 1992
58-0165	CMI	August 10, 1992
58-0176	CMI	November 17, 1992
58-0182	CMI	June 10, 1992
58-0231	CMI	December 3, 1992
59-2567	CMI	October 22, 1992
59-2568	CMI	November 24, 1992
59-2580	CMI	July 6, 1992
59-2581	CMI	October 8, 1992
59-2583	CMI	October 1, 1992
59-2590	CMI	July 13, 1992

B-52G 58-0164 is seen here in 379th Bomb Wing markings during September 1992. (Jerry Geer)

412th Test Wing, Edwards AFB, California

Serial Number	Weapon Configuration	Date Retired
58-0235	H I	March 29, 1994
59-2586	H I	April 22, 1994

BELOW: B-52G 58-0235 is seen here in 412th Test Wing markings at AMARC. (Author)

B-52H

ACC B-52H AIRCRAFT ASSIGNMENTS 1992-1998

SERIAL NUMBER	JUNE 1, 1992 ASSIGNMENT	TRANSFERS	CURRENT ASSIGNMENT
60-0001	PDM -Tinker AFB	PDM to Barksdale, Aug 92	20 BS Barksdale
60-0002	644 BS KI Sawyer	to 2 BW, Jun 94	11 BS Barksdale
60-0003	668 BS Griffiss	to KI, Jun 92 to 93 BS, AFRC, Apr 94	93 BS (AFRC) Barksdale
60-0004	23 BS Minot	none	23 BS Minot
60-0005	668 BS Griffiss	to KI, Jul 93 to 23 BS, Jul 94	23 BS Minot
60-0007	7 BW Carswell	to 20 BS, Barksdale, Dec18, 92 to 23 BS, Minot, Jun 95	23 BS Minot
60-0008	644 BS KI Sawyer	to 596 BS, Barksdale, May 93	96 BS Barksdale
60-0009	7 BW Carswell	to 596 BS, Barksdale, Dec 92 to 23 BS, Minot, Apr 96	23 BS Minot
60-0010	7 BW Carswell	to 596 BS, Barksdale, Nov 92	96 BS Barksdale
60-0011	PDM -Tinker AFB	to KI, Sep 92 to Barksdale, Sep 94	20 BS Barksdale
60-0012	668 BS Griffiss	to 11 BS, Barksdale, Nov 94 to 23 BS, Minot, Feb 96 to Barksdale, Jun 98	Minot (attrition reserve) On loan to 20 BS Barksdale
60-0013	PDM -Tinker AFB	to KI, Jul 92 to 23 BS, Minot, Oct 93 to Barksdale, May 94	96 BS Barksdale
60-0014	7BW Carswell	to Barksdale, Oct 92	20 BS Barksdale
60-0015	325 BS Fairchild	to 23 BS, Minot, Feb 94 23 BS Minot	(72 BS, Jan-Dec 95)
60-0016	7 BW Carswell	to 596 BS, Barksdale, Oct 92	96 BS Barksdale
60-0017	7 BW Carswell	to 20 BS, Barksdale, Dec 92	11 BS Barksdale
60-0018	7 BW Carswell	to 596 BS, Barksdale, Oct 92 to 23 BS, Minot, Apr 96	23 BS Minot
60-0019	7 BW Carswell	to Barksdale, Oct 92	11 BS Barksdale
60-0020	23 BS Minot	to Barksdale, Feb 93	20 BS Barksdale
60-0021	668 BS Griffiss	to KI, Sep 93 to Minot, Jul 94	23 BS Minot
60-0022	23 BS Minot	(72 BS, Jan-Dec 95) to Barksdale, Apr 96	96 BS Barksdale
60-0023	23 BS Minot	no moves	23 BS Minot
60-0024	668 BS Griffiss	to Minot, Nov 94	(72 BS, Jan-Dec 1995) 23 BS Minot
60-0025	7 BW Carswell	to Barksdale, Oct 92	20 BS Barksdale
60-0026	644 BS KI Sawyer	to Minot, Nov 18, 94	23 BS Minot
60-0028	325 BS Fairchild	to Griffiss, Apr 94 to Barksdale, Nov 94	96 BS Barksdale
60-0029	23 BS Minot	(72 BS, Jan-Dec 95)	23 BS Minot
60-0030	PDM -Tinker AFB	to Griffiss, Sep 92 to 20 BS, Barksdale, Oct 94	96 BS Barksdale
60-0031	7BW Carswell	to 96 BS, Barksdale, Dec 92	20 BS Barksdale
60-0032	7BW Carswell	to 20 BS, Barksdale, Oct 92	11 BS Barksdale
60-0033	668 BS Griffiss	to Minot, Oct 94	23 BS Minot
60-0034	644 BS KI Sawyer	to Minot, Nov 94	23 BS Minot
60-0035	644 BS KI Sawyer	to 20 BS, Barksdale, Aug 93	11 BS Barksdale
60-0036	644 BS KI Sawyer	to Minot, Nov 94 23 BS Minot	(72 BS, Jan-Dec 95)
60-0037	7BW Carswell	to Barksdale, Nov 92	96 BS Barksdale
60-0038	644 BS KI Sawyer	to Minot, Oct 94 to Barksdale, Feb 96	11 BS Barksdale

60-0041	325 BS Fairchild	to 2 BW, Barksdale, Feb 93	
		to 93 BS, AFRC, Mar 94	93 BS AFRC, Barksdale
60-0042	7BW Carswell	to Barksdale, Oct 92	
		to Minot, Mar 96	23 BS Minot
60-0043	644 BS KI Sawyer	to Barksdale, Dec 92	20 BS Barksdale
60-0044	325 BS Fairchild	to Minot, Apr 94	(72 BS, Jan-Dec 95)
		23 BS Minot	
60-0045	325 BS Fairchild	to 93 BS AFRC, Dec 7, 93	93 BS AFRC, Barksdale
60-0046	23 BS Minot	to 11 BS, Barksdale, Mar 96	11 BS Barksdale
60-0047	644 BS KI Sawyer	to Minot, Oct 94	(72 BS, Jan-Dec 95)
		23 BS Minot	
60-0048	644 BS KI Sawyer	to Barksdale, Oct 94	
		to Minot, Apr 96	
		to Barksdale (loan), Sep 97	Minot (attrition reserve)
			On loan to 11 BS Barksdale
60-0049	7BW Carswell	to Barksdale, Nov 92	20 BS Barksdale
60-0050	Edwards AFB	no moves	Edwards AFB
60-0051	7BW Carswell	to PDM, Aug 92	
		to Fairchild, Feb 93	
		to Minot, Apr 94	23 BS Minot
60-0052	7BW Carswell	to 325 BS Fairchild, Sep 92	
		to Minot, May 94	
		to Barksdale, Jun 95	96 BS Barksdale
60-0053	644 BS KI Sawyer	to Barksdale, Sep 93	96 BS Barksdale
60-0054	644 BS KI Sawyer	to Barksdale, Jul 94	96 BS Barksdale
60-0055	23 BS Minot	(72 BS, Jan-Dec 95)	23 BS Minot
60-0056	644 BS KI Sawyer	to Minot, Oct 94	23 BS Minot
60-0057	23 BS Minot	to Barksdale, Oct 94	20 BS Barksdale
60-0058	325 BS Fairchild	to Barksdale, Apr 94	
		to Minot, Sep 94	
		to Barksdale, Feb 96	11 BS Barksdale
60-0059	644 BS KI Sawyer	to Barksdale, Sep 93	96 BS Barksdale
60-0060	PDM -Tinker AFB	to Griffiss, Nov 92	
		to Minot, Nov 94	23 BS Minot
60-0061	23 BS Minot	(72 BS, Jan-Dec 95)	
		to 11 BS, Barksdale, Feb 96	11 BS Barksdale
60-0062	7BW Carswell	to Barksdale, Oct 92	20 BS Barksdale
60-0001	325 BS Fairchild	to Griffiss, Apr 94	
		to PDM, Oct 94	
		to 72 BS Minot, Mar 95	23 BS Minot
60-0002	7BW Carswell	to Barksdale, Nov 92	20 BS Barksdale
60-0003	325 BS Fairchild	to Minot, May 94	
		to Barksdale, Feb 96	11 BS Barksdale
60-0004	644 BS KI Sawyer	to Barksdale, Jun 94	20 BS Barksdale
60-0005	668 BS Griffiss	to KI, Sep 93	
		to Minot, Jul 94	(72 BS, Jan-Dec 95)
			23 BS Minot
60-0006	23 BS Minot	to Barksdale, Apr 96	11 BS Barksdale
60-0007	23 BS Minot		23 BS Minot
60-0008	325 BS Fairchild	to 93 BS AFRC, Dec 16, 93	93 BS AFRC, Barksdale
60-0009	325 BS Fairchild	to Barksdale, Apr 94	96 BS Barksdale
60-0010	23 BS Minot	to Barksdale, Apr 94	96 BS Barksdale
60-0011	644 BS KI Sawyer	to Barksdale, Aug 93	11 BS Barksdale
60-0012	668 BS Griffiss	to Barksdale, Nov 94	
		to Minot, Apr 96	
		to Barksdale (loan), Jan 98	Minot (attrition reserve)
			On loan to 11 BS Barksdale
60-0013	PDM -Tinker AFB	to Barksdale, Aug 92	
		to Minot, Feb 93	
		to Barksdale, Aug 94	96 BS Barksdale
60-0014	644 BS KI Sawyer	to Barksdale, May 94	
		to Minot, Mar 96	23 BS Minot
60-0015	644 BS KI Sawyer	to Griffiss, date unknown (1992?)	
		to Barksdale, Apr 94	
		to Minot, Feb 96	
		to Barksdale, Nov 97	Minot (attrition reserve)
			On loan to 11 BS Barksdale

60-0016	23 BS Minot	to Barksdale, Jun 94	11 BS Barksdale
60-0017	325 BS Fairchild	to 93 BS, AFRC, Jan 94	93 BS AFRC, Barksdale
60-0018	668 BS Griffiss	to KI, Aug 93 to Barksdale, May 94 to Minot, Apr 96	23 BS Minot
60-0019	644 BS KI Sawyer	to Barksdale, Jul 93	96 BS Barksdale
60-0020	644 BS KI Sawyer	to Griffiss, Nov 92 to Minot, Nov 94 to Barksdale, Apr 96	96 BS Barksdale
60-0021	PDM -Tinker AFB	to Griffiss, Dec 92 to 93 BS AFRC, Aug 94	93 BS AFRC, Barksdale
60-0022	23 BS Minot	to 93 BS AFRC, Feb 94	93 BS AFRC, Barksdale
60-0023	23 BS Minot	to Barksdale, Nov 94	20 BS Barksdale
60-0024	644 BS KI Sawyer	to Barksdale, Aug 94	96 BS Barksdale
60-0025	644 BS KI Sawyer	to Minot, 15 Nov 94 to Barksdale, Feb 98	Minot (attrition reserve) On loan to 11 BS Barksdale
60-0026	325 BS Fairchild	crashed, 24 Jun 94	
60-0027	7BW Carswell	to Barksdale, Oct 92 to Minot, Sep 95	23 BS Minot
60-0028	668 BS Griffiss	to Barksdale, Oct 92	96 BS Barksdale
60-0029	644 BS KI Sawyer	to Barksdale, Sep 93 to 93 BS AFRC, Apr 94	93 BS AFRC, Barksdale
60-0031	23 BS Minot	to Barksdale, Mar 96	11 BS Barksdale
60-0032	7BW Carswell	to Barksdale, Nov 92 to 93 BS AFRC, Jul 94	93 BS AFRC, Barksdale
60-0034	644 BS KI Sawyer	to Minot, Jul 94	23 BS Minot
60-0035	668 BS Griffiss	to KI, Aug 93 to Barksdale, May 94 to Minot, Feb 96	23 BS Minot
60-0036	325 BS Fairchild	to Minot, May 94	23 BS Minot
60-0038	7BW Carswell	to Barksdale, Oct 92	11 BS Barksdale
60-0039	23 BS Minot	(72 BS, Jan-Dec 95) to Barksdale, Feb 96	11 BS Barksdale (72 BS, Jan-Dec 95)
60-0040	325 BS Fairchild	to Minot, 26 Jun 94	23 BS Minot

ACC B-52H STRATOFORTRESS

The description of the B-52Hs include two data snapshots for each aircraft; one on June 30, 1993, and one on August 31, 1996. These data snapshots include the Unit/Base to which the aircraft was assigned and total flight hours at that time. Seventeen B-52Hs are presently kept as attrition reserves at Minot AFB.

B-52H-135-BW

60-0001 MEMPHIS BELLE IV - 2nd BOMB WING FLAGSHIP

60-0001, the first B-52H, was the 366th B-52 built at Boeing's Wichita, Kansas plant and was delivered to the Air Force on May 3, 1961. On June 30, 1993, 60-0001 was assigned to the 2nd Bomb Wing at Barksdale AFB and had accumulated 13,771.0 flight hours. By August 31, 1996 it was assigned to the 2nd Bomb Wing and had accumulated 14,985.4 flight hours. MEMPHIS BELLE IV visited the former Soviet Union twice after the end of the cold war, first at Poltava AB, Ukraine in September 1994 and second at Ryzan, Russia in May 1995. As of August 1, 1998, 60-0001 was marked as the 2nd Bomb Wing Flagship and assigned to the 20th Bomb Squadron at Barksdale AFB.

(Author)

BELOW: 60-0001 MEMPHIS BELLE IV, the 2nd Bomb Wing Flagship, is seen here on June 6, 1995, with MK 82 AIRs on a stub pylon and HSAB combination. (Brian C. Rogers)

60-0001 MEMPHIS BELLE IV, the 2nd Bomb Wing Flagship, is seen here on April 30, 1994, with AGM-86 missiles on a cruise missile pylon. (Brian C. Rogers)

60-0002

60-0002 was the 367th B-52 built at Boeing's Wichita, Kansas plant and was delivered to the Air Force on March 8, 1961. The aircraft was not modified to carry CSRL. On June 30, 1993, 60-0002 was assigned to the 2nd Bomb Wing at Barksdale AFB and had accumulated 12,479,2 flight hours. By August 31, 1996 it was assigned to the 2nd Bomb Wing and had accumulated 13.639.0 flight hours. As of August 1, 1998, 60-0002 was assigned to the 11th Bomb Squadron, 2nd Bomb Wing at Barksdale AFB.

60-0002 in 11th Bomb Squadron markings is seen here at Barksdale AFB on May 18, 1995. (Brian C. Rogers)

60-0002 is seen here at Barksdale AFB on May 5, 1995. (Author)

60-0003

The aircraft has received the Conventional Enhancement Modification (CEM). 60-0003 was the 368th B-52 built at Boeing's Wichita, Kansas plant and was delivered to the Air Force on March 3, 1961. On June 30, 1993, 60-0003 was assigned to the 410th Bomb Wing at K.I. Sawyer AFB and had accumulated 12,319.3 flight hours. By August 31, 1996 it was assigned to the 2nd Bomb Wing and had accumulated 13,110.7 flight hours. As of August 1, 1998, 60-0003 was assigned to the 93rd Bomb Squadron, 917th Wing, AFRC at Barksdale AFB.

60-0003 is seen here on May 5, 1995, at Barksdale AFB in 93rd Bomb Squadron markings, still wearing AFRES on the tail. (Author)

(Brian C. Rogers)

60-0004 THE BARON BUGLER

The aircraft has received CEM. 60-0004 was the 369th B-52 built at Boeing's Wichita, Kansas plant and was delivered to the Air Force on April 3, 1961. On June 30, 1993, 60-0004 was assigned to the 410th Bomb Wing at K.I Sawyer AFB and had accumulated 9,563.7 flight hours. By August 31, 1996 it was assigned to the 2nd Bomb Wing and had accumulated 10,424.2 flight hours. As of August 1, 1998, 60-0004 was assigned to the 23rd Bomb Squadron, 5th Bomb Wing at Minot AFB.

60-0004 is seen here at Minot AFB on April 1, 1998. (Brian C. Rogers)

60-0004 in 23rd Bomb Squadron markings is seen here on April 14, 1995, with MK 82 GP bombs on a stub pylon and MER beam. (Brian C. Rogers)

60-0005 WAR BIRDS
5th BOMB WING FLAGSHIP

The aircraft was not modified to carry CSRL. On June 30, 1993, 60-0005 was assigned to the 2nd Bomb Wing at Barksdale AFB and had accumulated 12,938.6 flight hours. By August 31, 1996 it was assigned to the 2nd Bomb Wing and had accumulated 13,894.6 flight hours. In addition to WAR BIRDS, it has been named WARLORD and also LOBO. As of August 1, 1998, 60-0005 was marked as the 5th Bomb Wing Flagship and assigned to the 23rd Bomb Squadron at Minot AFB as an attrition reserve aircraft.

(Author)

(Brian C. Rogers)

(Brian C. Rogers)

60-0005 is seen here at Minot AFB on February 24, 1996. It is nicknamed LOBO and marked as the 5th Bomb Wing Flagship. (Brian C. Rogers)

(Author)

60-0006 Crashed May 30, 1974 near Wright Patterson AFB, Ohio before change over to ACC

On May 30, 1974, 60-0006 crashed while assigned to 17th Bomb Wing due to loss of Stability Augmentation System (SAS) hydraulic and result of aircraft control.

60-0007 MEDAL OF HONOR NOSE ART

The aircraft has received CEM. 60-0007 was the 7th Bomb Wing Flagship, and the only B-52 to carry the CW (**C**ars**W**ell) tail code. With the inactivation of the 7th Bomb Wing at Carswell, 60-0007 moved to Barksdale AFB. On June 30, 1993, 60-0007 was assigned to the 2nd Bomb Wing at Barksdale AFB and had accumulated 12,653.8 flight hours. It was the 7th Bomb Wing's last B-52, leaving Carswell in December 1992 after the 7th Bomb Wing had operated B-52 for over 34 years. By August 31, 1996 it was assigned to the 5th Bomb Wing and had accumulated 13,661.2 flight hours. As of August 1, 1998, 60-0007 was assigned to the 23rd Bomb Squadron, 5th Bomb Wing at Minot AFB. While assigned to the 23rd BS, 60-0007 took part in Operation Desert Fox, launching CALCMs

LEFT: 60-0007 is seen here at Barksdale AFB on May 2, 1994, in the markings of the 20th Bomb Squadron. (Author)

60-0007 was photographed on November 30, 1992 at Carswell wearing a CW tail code. (Randy Walker)

60-0007 is seen here at Minot AFB on August 13, 1996, in 23rd Bomb Squadron markings. It is carrying MK 82 AIRs on the AGM-28 pylon and MER beam. (Brian C. Rogers)

60-0008 LUCKY LADY IV
8th AIR FORCE FLAGSHIP

Initially assigned to the 410th Bomb Wing at K.I. Sawyer, 60-0008 was transferred to Barksdale. On June 30, 1993, 60-0008 was assigned to the 2nd Bomb Wing at Barksdale AFB and had accumulated 12,994.1 flight hours. On August 1 thru 3, 1994 LADY LUCK IV took part in a 47 hour around the world flight. By August 31, 1996 it was assigned to the 2nd Bomb Wing and had accumulated 14,084.0 flight hours. As of August 1, 1998, 60-0008 was marked as the 8th Air Force Flagship and was assigned to the 96th Bomb Squadron, 2nd Bomb Wing at Barksdale AFB.

(Brian C. Rogers)

LEFT: 60-0008 is seen here at Barksdale AFB on June 3 1997, in 96th Bomb Squadron markings and marked as the 8th AF Flagship. (Author)

60-0008 is seen here at Ellsworth AFB in July 1992, in 410th Bomb Wing markings. (Author)

60-0008 is seen here at Barksdale AFB on April 6, 1998, in 96th Bomb Squadron markings and marked as the 8th AF Flagship. (Brian C. Rogers)

ROLLING THUNDER II nose art was applied to 60-0009 as seen in this photo taken on June 30, 1996. (Kevin Patrick)

60-0009 ROLLING THUNDER II

The aircraft has received CEM. 60-0009 was the 9th Bomb Squadron Flagship at Carswell AFB. With the inactivation of the 7th Bomb Wing at Carswell, 60-0009 moved to Barksdale AFB. On June 30, 1993, 60-0009 was assigned to the 2nd Bomb Wing at Barksdale AFB and had accumulated 13,318.3 flight hours. By August 31, 1996 it was assigned to the 5th Bomb Wing. As of August 1, 1998, 60-0009 was assigned to the 23rd Bomb Squadron, 5th Bomb Wing at Minot AFB.

60-0009 is seen here in 96th Bomb Squadron markings at Ellsworth AFB on June 21, 1993, as one of the B-52s deployed to support flying activities of the Bomber Weapons School. (Author)

60-0010 INSTIGATOR

The aircraft has received CEM. On June 30, 1993, 60-0010 was assigned to the 2nd Bomb Wing at Barksdale AFB and had accumulated 13,919.0 flight hours. By August 31, 1996 it was assigned to the 2nd Bomb Wing and had accumulated 14,673.1 flight hours. It was nicknamed INSTIGATOR. As of August 1, 1998, 60-0010 was assigned to the 96th Bomb Squadron, 2nd Bomb Wing at Barksdale AFB. 60-0010 took part in Operation Desert Fox, flying CALCM missions from Diego Garcia.

(Author)

60-0010 is seen here at Barksdale AFB on May 5, 1995. (Author)

(Author)

60-0011

The aircraft has received CEM and AGM-142 Have Nap capability. On June 30, 1993, 60-0011 was assigned to the 2nd Bomb Wing at Barksdale AFB and had accumulated 13,137.3 flight hours. By August 31, 1996 it was assigned to the 2nd Bomb Wing and had accumulated 14,379.4 flight hours. As of August 1, 1998, 60-0011 was assigned to the 20th Bomb Squadron, 2nd Bomb Wing at Barksdale AFB.

60-0011 is seen here at Offutt AFB during August 1993. (George Cockle)

60-0012

The aircraft was not modified to carry CSRL. Since transitioning to ACC 60-0012 has been assigned to the 92nd Bomb Wing, 416th Bomb Wing, and the 2nd Bomb Wing. On June 30, 1993, 60-0012 was assigned to the 416th Bomb Wing at Griffiss AFB and had accumulated 15,561.7 flight hours. By August 31, 1996 it was assigned to the 2nd Bomb Wing and had accumulated 16,492.0 flight hours. As of August 1, 1998, 60-0012 was assigned to the 96th Bomb Squadron, 2nd Bomb Wing at Barksdale AFB, loaned from Minot's attrition reserve.

BELOW: 60-0012 seen here at Ellsworth on July 27, 1992, taxiing back from a mission. It is still in the old SAC 416th Bomb Wing markings. Of interest is the white frost that has formed on the external tank. The frost is caused by the fuel which became very cold during the bomber's flight at high altitude. (Author)

60-0012 is seen here at Griffiss AFB in April, 1994, with the 416th Bomb Wing's ACC markings. (Paul Hart)

60-0012 is seen here at Barksdale AFB on May 5, 1995, in 11th Bomb Squadron markings. (Author)

60-0013

The aircraft was one of the four B-52Hs to receive Rapid AGM-84 Harpoon. On June 30, 1993, 60-0013 was assigned to the 410th Bomb Wing at K.I. Sawyer AFB and had accumulated 13,480.1 flight hours. By August 31, 1996 it was assigned to the 2nd Bomb Wing and had accumulated 14,724.7 flight hours. As of August 1, 1998, 60-0013 was assigned to the 96th Bomb Squadron, 2nd Bomb Wing at Barksdale AFB.

60-0013 is seen here at Barksdale AFB on May 18, 1995, in 96th Bomb Squadron markings. (Brian C. Rogers)

B-52H -140 -BW

60-0014 GLOBAL REACH & POWER

The aircraft has received CEM and was one of the four B-52Hs to receive the Rapid Eight AGM-142 Have Nap capability. It was also modified to the standard AGM-142 Have Nap capability. On June 30, 1993, 60-0014 was assigned to the 2nd Bomb Wing at Barksdale AFB and had accumulated 13,597.3 flight hours. By August 31, 1996 it was assigned to the 2nd Bomb Wing and had accumulated 14,614.9 flight hours. As of August 1, 1998, 60-0014 was assigned to the 20th Bomb Squadron, 2nd Bomb Wing at Barksdale AFB.

As DUKE 02 the aircraft was the number two aircraft of the CALCM attack against Iraq on September 3, 1996, launching six of eight CALCMs from the internal CSRL. The targets were command and control facilities and radar sites.

(Brian C. Rogers)

60-0014 is seen here at Barksdale AFB on June 5, 1995, in 20th Bomb Squadron markings. (Brian C. Rogers)

60-0015 is seen here at Minot AFB on February 24, 1996, in 23rd Bomb Squadron markings. (Brian C. Rogers)

60-0015

The aircraft has received CEM. On June 30, 1993, 60-0015 was assigned to the 92nd Bomb Wing at Fairchild AFB and had accumulated 13,241.1 flight hours. By August 31, 1996 it was assigned to the 2nd Bomb Wing and had accumulated 14,183.9 flight hours. As of August 1, 1998, 60-0015 was assigned to the 23rd Bomb Squadron 5th Bomb Wing at Minot. 60-0015 took part in Operation Desert Fox.

60-0015 is seen here at Minot AFB on April 26, 1995, with the 5th Bomb Wing red and white candy stripe tail cap. (Brian C. Rogers)

60-0016

The aircraft has received CEM. On June 30, 1993, 60-0016 was assigned to the 2nd Bomb Wing at Barksdale AFB and had accumulated 14,389.2 flight hours. By August 31, 1996 it was assigned to the 2nd Bomb Wing and had accumulated 15,466.1 flight hours. As of August 1, 1998, 60-0016 was assigned to the 96th Bomb Squadron, 2nd Bomb Wing at Barksdale AFB. 60-0016 took part in Operation Desert Fox.

60-0016 is seen here at Barksdale AFB on May 5, 1995, in 96th Bomb Squadron markings. (Author)

60-0017

On June 30, 1993, 60-0017 was assigned to the 2nd Bomb Wing at Barksdale AFB and had accumulated 14,606.4 flight hours. By August 31, 1996 it was assigned to the 2nd Bomb Wing and had accumulated 15,704.5 flight hours. As of August 1, 1998, 60-0017 was assigned to the 11th Bomb Squadron, 2nd Bomb Wing at Barksdale AFB, and was at Tinker AFB undergoing Programmed Depot Maintenance and addition of CEM.

60-0012 is seen here at Barksdale AFB on April 6, 1998, in 11th Bomb Squadron markings. (Brian C. Rogers)

60-0012 is seen here at Barksdale AFB on May 5, 1995, in 20th Bomb Squadron markings. (Author)

(Brian C. Rogers)

60-0018 GUIDED SPIRIT

The aircraft has received CEM. On June 30, 1993, 60-0018 was assigned to the 2nd Bomb Wing at Barksdale AFB and had accumulated 13,810.4 flight hours. By August 31, 1996 it was assigned to the 5th Bomb Wing and had accumulated 14,705.8 flight hours. It was nicknamed DARTH GATOR at Barksdale and renamed GUIDED SPIRIT when reassigned to Minot. As of August 1, 1998, 60-0018 was assigned to the 23rd Bomb Squadron, 5th Bomb Wing at Minot AFB.

(Brian C. Rogers)

60-0018 is seen here at Minot AFB on May 18, 1996, in 23rd Bomb Squadron markings. (Brian C. Rogers)

60-0018 is seen here at Tinker on August 5, 1995, during PDM (Programmed Depot Maintenance). The shiny areas on the fuselage are new aluminum skins. (Author)

60-0019

On June 30, 1993, 60-0019 was assigned to the 2nd Bomb Wing at Barksdale AFB and had accumulated 13,229.9 flight hours. By August 31, 1996 it was assigned to the 2nd Bomb Wing and had accumulated 14,140.8 flight hours. As of August 1, 1998, 60-0019 was assigned to the 11th Bomb Squadron, 2nd Bomb Wing at Barksdale AFB, but was at Tinker AFB undergoing Programmed Depot Maintenance and addition of CEM.

BELOW: 60-0019 is seen here at Barksdale AFB on May 5, 1995, in 11th Bomb Squadron markings. (Author)

60-0020 MAD BOLSHEVIK
20th BOMB SQUADRON FLAGSHIP

The aircraft has received CEM and AGM-142 Have Nap capability. On June 30, 1993, 60-0020 was assigned to the 2nd Bomb Wing at Barksdale AFB and had accumulated 12,660.2 flight hours. By August 31, 1996 it was assigned to the 2nd Bomb Wing and had accumulated 13,811.9 flight hours. As of August 1, 1998, 60-0020 was marked as the 20th Bomb Squadron Flagship and assigned to the 2nd Bomb Wing at Barksdale AFB.

(Author)

LEFT: (Brian C. Rogers)

BELOW: 60-0020 THE MAD BOLSHEVIK, the 20th Bomb Squadron Flagship, is seen here at Barksdale AFB on June 30, 1997. (Author)

60-0021

The aircraft was not modified to carry CSRL. 60-0021 was one of the two B-52Hs taking part in the Senior Bowl program as one of the D-21 reconnaissance drone launchers. On June 30, 1993, 60-0021 was assigned to the 410th Bomb Wing at K.I. Sawyer AFB and had accumulated 13,828.6 flight hours. By August 31, 1996 it was assigned to the 5th Bomb Wing and had accumulated 14,608.8 flight hours. As of August 1, 1998, 60-0021 was assigned to the 23rd Bomb Squadron, 5th Bomb Wing at Minot AFB as an attrition reserve aircraft.

60-0021 is seen here at Minot AFB on September 21, 1995, in 23rd Bomb Squadron markings. (Brian C. Rogers)

60-0021 is seen here at Ellsworth AFB on June 30, 1997, in 23rd Bomb Squadron markings. (Author)

B-52H -145 -BW

60-0022
The aircraft has received CEM. On June 30, 1993, 60-0022 was assigned to the 5th Bomb Wing at Minot AFB and had accumulated 12,030.8 flight hours. By August 31, 1996 it was assigned to the 2nd Bomb Wing and had accumulated 12,965.5 flight hours. As of August 1, 1998, 60-0022 was assigned to the 96th Bomb Squadron, 2nd Bomb Wing at Barksdale AFB.

60-0022 is seen here at Barksdale AFB on May 5, 1995, in 96th Bomb Squadron markings. (Author)

60-0022 is seen here at Minot AFB on August 15, 1995, with the 5th Bomb Wing red and white candy cane stripe tail cap. (Brian C. Rogers)

60-0023 BOMBER BARONS
23rd BOMB SQUADRON FLAGSHIP

The aircraft has received CEM. On June 30, 1993, 60-0023 was assigned to the 92nd Bomb Wing at Fairchild AFB and had accumulated 12,810.5 flight hours. By August 31, 1996 it was assigned to the 2nd Bomb Wing and had accumulated 13,462.6 flight hours. As of August 1, 1998, 60-0023 was marked as the 23rd Bomb Squadron Flagship and assigned to the 5th Bomb Wing at Minot AFB. 60-0023 took part in Operation Desert Fox.

(Brian C. Rogers)

(Brian C. Rogers)

60-0023 BOMBER BARONS, the 23rd Bomb Squadron Flagship, is seen here running up engines for takeoff at Minot AFB on May 13, 1996. (Brian C. Rogers)

60-0024
The aircraft was not modified to carry CSRL. On June 30, 1993, 60-0024 was assigned to the 416th Bomb Wing at Griffiss AFB and had accumulated 14,974.2 flight hours. While at Griffiss 60-0024 was nicknamed GUARDIAN and was one of the last two B-52s to leave Griffiss, taking part in a ceremonial flyby on November 15, 1994. By August 31, 1996, it was assigned to the 5th Bomb Wing and had accumulated 15,928.3 flight hours. As of August 1, 1998, 60-0024 was assigned to the 23rd Bomb Squadron, 5th Bomb Wing at Minot AFB as an attrition reserve aircraft.

(Brian C. Rogers)

LEFT: 60-0024 is seen here at Minot AFB on August 15, 1995, with the 5th Bomb Wing red and white candy stripe tail cap. (Brian C. Rogers)

60-0024 is seen here at Nellis AFB on October 14, 1993, in markings of the 416th Bomb Wing. (Brian C. Rogers)

60-0022 is seen here at Minot AFB on June 23, 1995, with the black 72nd Bomb Squadron tail cap. (Brian C. Rogers)

(Brian C. Rogers)

60-0025 OL CROW EXPRESS II

The aircraft has received CEM and was one of the four B-52Hs to receive Rapid Eight AGM-142 Have Nap capability. It was also modified to the standard AGM-142 Have Nap capability. On June 30, 1993, 60-0025 was assigned to the 2nd Bomb Wing at Barksdale AFB and had accumulated 12,722.5 flight hours. By August 31, 1996 it was assigned to the 2nd Bomb Wing and had accumulated 13,658.6 flight hours. As of August 1, 1998, 60-0025 was assigned to the 20th Bomb Squadron, 2nd Bomb Wing at Barksdale AFB.

As DUKE 03 the aircraft was the spare aircraft on the CALCM attack against Iraq on September 3, 1996. With 60-0059 and 60-0014 functioning properly 60-0025 was not needed for the strike and returned to base without launching any CALCMS. It returned to the Gulf during Operation Desert Fox launching CALCMs against Iraq.

60-0025 OL CROW EXPRESS II is seen here at Barksdale AFB on June 30, 1997, in 20th Bomb Squadron Markings. (Author)

60-0025 is seen here at Barksdale AFB on May 5, 1995, carrying an AGM-142 CATM (Captive Air Training Missile). (Author)

60-0026 PREDATOR

The aircraft has received CEM. On June 30, 1993, 60-0026 was assigned to the 410th Bomb Wing at K.I. Sawyer AFB and had accumulated 12,705.2 flight hours. By August 31, 1996 it was assigned to the 5th Bomb Wing and had accumulated 13,593.7 flight hours. As of August 1, 1998, 60-0026 was assigned to the 23rd Bomb Squadron, 5th Bomb Wing at Minot AFB.

(Brian C. Rogers)

60-0026 PREDATOR is seen here at Minot AFB on April 14, 1996, in the markings of the 23rd Bomb Squadron, 5th Bomb Wing. (Brian C. Rogers)

60-0026 is seen here on July 25, 1994, in the markings of the 410th Bomb Wing from K.I. Sawyer AFB, Michigan. (Norris Graser)

60-0027 Crashed on October 4, 1968 near Minot AFB, North Dakota before change over to ACC
The aircraft crashed due to fuel mismanagement, four were killed and two survived.

60-0028
On June 30, 1993, 60-0028 was assigned to the 92nd Bomb Wing at Fairchild AFB and had accumulated 12,978.9 flight hours. By August 31, 1996 it was assigned to the 2nd Bomb Wing and had accumulated 14,041.4 flight hours. As of August 1, 1998, 60-0028 was assigned to the 96th Bomb Squadron, 2nd Bomb Wing at Barksdale AFB.

60-0028 is seen here at Fairchild AFB on August 7, 1993, with FC tail codes and the checkerboard tail cap of the 92nd Bomb Wing. (R.E.F. Jones)

60-0028 is seen here at Barksdale AFB on May 5, 1995, in the markings of the 96th Bomb Squadron, 2nd Bomb Wing. (Author)

60-0029

On June 30, 1993, 60-0029 was assigned to the 5th Bomb Wing at Minot AFB and had accumulated 11,713.1 flight hours. By August 31, 1996 it was assigned to the 2nd Bomb Wing and had accumulated 12,469.3 flight hours. As of August 1, 1998, 60-0029 was assigned to the 23rd Bomb Squadron, 5th Bomb Wing at Minot AFB as an attrition reserve aircraft.

60-0029 is seen here at in 5th Bomb Wing markings at Ellsworth AFB on June 21, 1993, as one of the B-52s deployed to support flying activities of the Bomber Weapons School. (Author)

60-0029 is seen here at Minot AFB on March 28, 1996, in the markings of the 23rd Bomb Squadron, 5th Bomb Wing. (Brian C. Rogers)

60-0029 is seen here at Minot AFB on April 14, 1996, in the markings of the 72nd Bomb Squadron, 5th Bomb Wing. (Brian C. Rogers)

(Brian C. Rogers)

60-0030

On June 30, 1993, 60-0030 was assigned to the 416th Bomb Wing at Griffiss AFB and had accumulated 13,075.3 flight hours. While at Griffiss, 60-0030 was nicknamed MISS LIBERTY and marked as the 668th Bomb Squadron Flagship. By August 31, 1996 it was assigned to the 2nd Bomb Wing and had accumulated 14,192.4 flight hours. As of August 1, 1998, 60-0030 was assigned to the 20th Bomb Squadron, 2nd Bomb Wing at Barksdale AFB.

60-0030 is seen here at Barksdale AFB on June 30, 1997, in the markings of the 20th Bomb Squadron, 2nd Bomb Wing. (Author)

60-0030 is seen here at Barksdale AFB on June 14, 1995, in the markings of the 96th Bomb Squadron, 2nd Bomb Wing. (Brian C. Rogers)

60-0030 MISS LIBERTY, the 668th Bomb Squadron Flagship, is seen here at Nellis AFB on October 14, 1993. (Brian C. Rogers)

60-0031

The aircraft has received CEM and AGM-142 Have Nap capability. On June 30, 1993, 60-0031 was assigned to the 2nd Bomb Wing at Barksdale AFB and had accumulated 13,382.9 flight hours. By August 31, 1996 it was assigned to the 2nd Bomb Wing and had accumulated 14,182.3 flight hours. As of August 1, 1998, 60-0031 was assigned to the 20th Bomb Squadron, 2nd Bomb Wing at Barksdale AFB.

60-0031 is seen here in 96th Bomb Squadron, 5th Bomb Wing markings at Ellsworth AFB on June 21, 1993, as one of the B-52s deployed to support flying activities of the Bomber Weapons School. (Author)

60-0031 is seen here at Barksdale AFB on May 5, 1995, in the markings of the 20th Bomb Squadron, 2nd Bomb Wing. (Author)

60-0032

The aircraft has received CEM. On June 30, 1993, 60-0032 was assigned to the 2nd Bomb Wing at Barksdale AFB and had accumulated 12,108.9 flight hours. By August 31, 1996 it was assigned to the 2nd Bomb Wing and had accumulated 14,294.9 flight hours. As of August 1, 1998, 60-0032 was assigned to the 11th Bomb Squadron, 2nd Bomb Wing at Barksdale AFB.

60-0032 is seen here at Barksdale AFB on May 8, 1996, in the markings of the 20th Bomb Squadron, 2nd Bomb Wing. (Author)

60-0032 is seen here in 96th Bomb Squadron, 5th Bomb Wing markings at Ellsworth AFB on June 21, 1993, as one of the B-52s deployed to support flying activities of the Bomber Weapons School. (Author)

60-0033 INSTRUMENT OF DESTRUCTION

The aircraft has received CEM. On June 30, 1993, 60-0033 was assigned to the 416th Bomb Wing at Griffiss AFB and had accumulated 14,719.3 flight hours. While at Griffiss, 60-0033 was nicknamed FRANCE LIBRE and marked as the 416th Bomb Wing Flagship. By August 31, 1996 it was assigned to the 5th Bomb Wing and had accumulated 15,482.5 flight hours. As of August 1, 1998, 60-0033 was nicknamed INSTRUMENT OF DESTRUCTION and assigned to the 23rd Bomb Squadron, 5th Bomb Wing at Minot AFB.

(Brian C. Rogers)

(Randy Walker)

60-0033 FRANCE LIBRE, the 416th Bomb Wing Flagship is seen here at Tinker AFB on March 27, 1993. (Randy Walker)

60-0033 INSTRUMENT OF DESTRUCTION is seen here at Minot AFB on March 9, 1996, in markings of the 23rd Bomb Squadron, 5th Bomb Wing. (Brian C. Rogers)

(Brian C. Rogers)

B-52H -150 -BW

60-0034 WISE GUY

The aircraft has received CEM. On June 30, 1993, 60-0034 was assigned to the 410th Bomb Wing at K.I. Sawyer AFB and had accumulated 13,879.1 flight hours. By August 31, 1996 it was assigned to the 5th Bomb Wing and had accumulated 14,520.9 flight hours. It carries the nickname WISE GUY with associated nose art. As of August 1, 1998, 60-0034 was assigned to the 23rd Bomb Squadron, 5th Bomb Wing at Minot AFB. 60-0034 took part in Operation Desert Fox, flying CALCM missions from Diego Garcia.

60-0034 WISE GUY is seen here at Nellis AFB on October 11, 1996, in 23rd Bomb Squadron markings. (Brian C. Rogers)

60-0035

On June 30, 1993, 60-0035 was assigned to the 410th Bomb Wing at K.I. Sawyer AFB and had accumulated 12,922.1 flight hours. By August 31, 1996 it was assigned to the 2nd Bomb Wing and had accumulated 13,997.4 flight hours. As of August 1, 1998, 60-0035 was assigned to the 11th Bomb Squadron, 2nd Bomb Wing at Barksdale AFB.

60-0035 is seen here at Barksdale AFB on May 5, 1995, in the markings of the 20th Bomb Squadron, 2nd Bomb Wing. (Author)

60-0035 is seen here on May 31, 1996, in the markings of the 11th Bomb Squadron, 2nd Bomb Wing. (Tom Kaminski)

(Brian C. Rogers)

60-0036 TAGBOARD FLYER

On June 30, 1993, 60-0036 was assigned to the 410th Bomb Wing at K.I. Sawyer AFB and had accumulated 12,601.0 flight hours. By August 31, 1996 it was assigned to the 5th Bomb Wing and had accumulated 13,365.6 flight hours. In recognition of its part in the Senior Bowl program as one of the D-21 reconnaissance drone launchers, it carries the nickname TAGBOARD FLYER and associated nose art. As of August 1, 1998, 60-0036 was assigned to the 23rd Bomb Squadron, 5th Bomb Wing at Minot AFB as an attrition reserve aircraft. 60-0036 took part in Opertion Desert Fox.

60-0036 TAGBOARD FLYER is seen here at Minot AFB on May 14, 1997, in the markings of the 23rd Bomb Squadron, 5th Bomb Wing. (Brian C. Rogers)

60-0036 is seen here at Minot AFB on November 5, 1995, in the markings of the 72nd Bomb Squadron, 5th Bomb Wing, with AGM-129A ACMs on the wing pylon. (Brian C. Rogers)

60-0037 WHAM BAM II

The aircraft has received CEM. On June 30, 1993, 60-0037 was assigned to the 2nd Bomb Wing at Barksdale AFB and had accumulated 12,726.3 flight hours. By August 31, 1996 it was assigned to the 2nd Bomb Wing and had accumulated 13,626.5 flight hours. During late July 1998 nose art of an early representation of Bugs Bunny was added to 60-0037. This art named WHAM BAM II had originally been on a B-24 (41-23738). The original WHAM BAM flew in Europe with the 453rd Bomb Group, and flown by Captain Jimmie Stewart. The B-24 was retired following the war and scrapped. As of August 1, 1998, 60-0037 was assigned to the 96th Bomb Squadron, 2nd Bomb Wing at Barksdale AFB.

60-0037 is seen here at Barksdale AFB on June 30, 1997, in the markings of the 96th Bomb Squadron, 2nd Bomb Wing. (Author)

60-0038

On June 30, 1993, 60-0038 was assigned to the 410th Bomb Wing at K.I. Sawyer AFB and had accumulated 12,715.8 flight hours. By August 31, 1996 it was assigned to the 2nd Bomb Wing and had accumulated 13,886.8 flight hours. As of August 1, 1998, 60-0038 was assigned to the 11th Bomb Squadron, 2nd Bomb Wing at Barksdale AFB.

RIGHT: 60-0038 is seen here at Minot AFB on September 15, 1995, in the markings of the 5th Bomb Wing. It is carrying MK 82 AIR 500 pound bombs on the AGM-28 pylon and MER rack combination.(Brian C. Rogers)

60-0038 is seen here in May 1993, in 410th Bomb Wing markings with AGM-126 Advanced Cruise Missiles (ACM) on the Wing pylons. (David Brown)

60-0039 Crashed on April 1, 1977 at K.I. Sawyer AFB, Michigan before change over to ACC

On April 1, 1977, 60-0039 was assigned to the 410th Bomb Wing at K.I. Sawyer. The aircraft crashed on landing approach. All eight crew members on board were killed

60-0040 Crashed on December 6, 1988 at K.I. Sawyer AFB, Michigan before change over to ACC

On December 6, 1988, 60-0040 was assigned to the 410th Bomb Wing at K.I. Sawyer. The aircraft exploded during a touch and go landing.

60-0041

The aircraft has received CEM. On June 30, 1993, 60-0041 was assigned to the 2nd Bomb Wing at Barksdale AFB and had accumulated 14,542.7 flight hours. By August 31, 1996 it was assigned to the 917th Wing and had accumulated 15,291.1 flight hours. As of August 1, 1998, 60-0041 was assigned to the 93rd Bomb Squadron, 917th Wing, AFRC at Barksdale AFB.

BELOW: 60-0041 is seen here at Barksdale AFB on June 30, 1997, in the markings of the 93rd Bomb Squadron, 917th Wing, AFRES. (Author)

60-0042 is seen here at Minot AFB in February 8, 1997, in 23rd Bomb Squadron, 5th Bomb Wing markings. (Brian C. Rogers)

60-0042

60-0042 was the first B-52-H to arrive at Barksdale AFB during the B-52G to B-52H conversion, arriving on August 20, 1992. On June 30, 1993 60-0042 still assigned to the 2nd Bomb Wing at Barksdale AFB, had accumulated 13,120.7 flight hours. By August 31, 1996 it was assigned to the 5th Bomb Wing and had accumulated 13,919.7 flight hours. As of August 1, 1998, 60-0042 was assigned to the 23rd Bomb Squadron, 5th Bomb Wing at Minot AFB as an attrition reserve aircraft.

60-0042 is seen here at Barksdale AFB on June 5, 1995, taxiing in 11th Bomb Squadron Markings. (Brian C. Rogers)

60-0043

The aircraft has received CEM and AGM-142 Have Nap capability. On June 30, 1993, 60-0043 was assigned to the 2nd Bomb Wing at Barksdale AFB and had accumulated 12,934.1 flight hours. It was the first CEM aircraft at Barksdale AFB arriving in December 1994. As part of a Global Power exercise, Bright Star 95, 60-0043 flew on November 14, 1995 a nonstop mission from Barksdale AFB to Egypt and returned to Barksdale. By August 31, 1996 it was assigned to the 2nd Bomb Wing and had accumulated 14,255.3 flight hours. As of August 1, 1998, 60-0043 was assigned to the 20th Bomb Squadron, 2nd Bomb Wing at Barksdale AFB. 60-0043 took part in Operation Desert Fox.

60-0043 is seen here at Barksdale AFB on May 5, 1995, in 20th Bomb Squadron markings carrying an AGM-142 CATM (Captive Air Training Missile). (Author)

60-0043 is seen here at Barksdale AFB on October 10, 1997, in 20th Bomb Squadron markings. (Brian C. Rogers)

(Brian C. Rogers)

60-0044 MAGNUM FORCE

The aircraft has received CEM. On June 30, 1993, 60-0044 was assigned to the 92nd Bomb Wing at Fairchild AFB and had accumulated 13,578.9 flight hours. By August 31, 1996 it was assigned to the 5th Bomb Wing and had accumulated 14,445.0 flight hours. As of August 1, 1998, 60-0044 it carried the nickname MAGNUM FORCE and associated nose art, and was assigned to the 23rd Bomb Squadron, 5th Bomb Wing at Minot AFB. 60-0044 took part in Operation Desert Fox.

60-0044 is seen here at Minot AFB on August 13, 1996, in 23rd Bomb Squadron markings carrying MK 82 AIR 500 pound bombs on an AGM-28 pylon, MER beam combination. (Brian C. Rogers)

60-0044 is seen here in 72nd Bomb Squadron markings on December 10, 1995, during a Global Power mission. (Brian C. Rogers)

60-0045

On June 30, 1993, 60-0045 was assigned to the 92nd Bomb Wing at Fairchild AFB and had accumulated 13,085.0 flight hours. By August 31, 1996 it was assigned to the 2nd Bomb Wing and had accumulated 13,694.0 flight hours. 60-0045 was the first B-52H, also the first strategic bomber, to be assigned to the Air Force Reserve. As of August 1, 1998, 60-0045 was assigned to the 93rd Bomb Squadron, 917th Wing, AFRC at Barksdale AFB.

(Author)

60-0045 is seen here at Barksdale AFB on May 8, 1996, in the markings of the 93rd Bomb Squadron, 917th Bomb Wing, AFRES. (Author)

60-0045 is seen here at Barksdale AFB on June 30, 1997, in the markings of the 93rd Bomb Squadron, 917th Bomb Wing, AFRC. (Author)

B-52H -155 -BW

60-0046
The aircraft has received CEM. On June 30, 1993, 60-0046 was assigned to the 5th Bomb Wing at Minot AFB and had accumulated 11,799.1 flight hours. By August 31, 1996 it was assigned to the 2nd Bomb Wing and had accumulated 13,402.4 flight hours. As of August 1, 1998, 60-0046 was assigned to the 11th Bomb Squadron, 2nd Bomb Wing at Barksdale AFB. 60-0046 flew Operation Desert Fox missions from Diego Garcia launching CALCMs against Iraq.

60-0046 is seen here at Minot AFB on April 14, 1995, in the red and white candy stripe markings of the 5th Bomb Wing. (Author)

60-0047 NORTH DAKOTA PAYBACK

On June 30, 1993, 60-0047 was assigned to the 410th Bomb Wing at K.I. Sawyer AFB and had accumulated 13,484.2 flight hours. On May 5, 1995, while assigned to the 72nd Bomb Squadron, 60-0047 suffered a double engine failure (engines 3 and 4) and engine explosion. The vertical fin was badly damaged. It was replaced with the fin from B-52G 59-2570 and returned to flying status. By August 31, 1996 it was assigned to the 5th Bomb Wing and had accumulated 14,349.3 flight hours. As of August 1, 1998, 60-0047 was assigned to the 23rd Bomb Squadron, 5th Bomb Wing at Minot AFB as an attrition reserve aircraft.

(Brian C. Rogers)

60-0047 is seen here at Minot AFB on March 28, 1996, in the black tail stripe of the 72nd Bomb Squadron, 5th Bomb Wing. (Brian C. Rogers)

60-0047 is seen here at Minot AFB on August 18, 1995, with the vertical fin from B-52G 59-2570. The original vertical fin was heavily damaged as a result of a catastrophic engine failure. (Brian C. Rogers)

60-0048

The aircraft was not modified to carry CSRL. On June 30, 1993, 60-0048 was assigned to the 416th Bomb Wing at Griffiss AFB and had accumulated 13,584.6 flight hours. By August 31, 1996 it was assigned to the 5th Bomb Wing and had accumulated 14,823.3 flight hours. As of August 1, 1998, 60-0048 was assigned to the 11th Bomb Squadron, 2nd Bomb Wing at Barksdale AFB coming from the attrition reserve at Minot AFB.

ABOVE: 60-0048 is seen here at Barksdale AFB on May 15, 1995, with the yellow tail stripe of the 11th Bomb Squadron, 2nd Bomb Wing. (Brian C. Rogers) BELOW: 60-0048 is seen here at Griffiss AFB on April 1 1994, in the markings of the 668th Bomb Squadron, 416th Bomb Wing. (Tom Kaminski)

60-0049

The aircraft has received CEM. On June 30, 1993, 60-0049 was assigned to the 2nd Bomb Wing at Barksdale AFB and had accumulated 13,836.6 flight hours. By August 31, 1996 it was assigned to the 2nd Bomb Wing and had accumulated 14,892.5 flight hours. As of August 1, 1993, 60-0049 was assigned to the 20th Bomb Squadron, 2nd Bomb Wing at Barksdale AFB.

60-0049 is seen here at Barksdale AFB on April 30, 1994, with an AGM-28 pylon and an empty MER beam under the Wing. (Brian C. Rogers)

60-0050 DRAGON'S INFERNO

The aircraft has received CEM and AGM-142 Have Nap capability. 60-0050 is the single B-52H permanently assigned to test duties at Edwards AFB. On June 30, 1993, 60-0050 was assigned to the 412th Test Wing at Edwards AFB and had accumulated 11,742.1 flight hours. By August 31, 1996 it was assigned to the 412th Test Wing and had accumulated 11,940.4 flight hours. As of August 1, 1998, 60-0050 carried the nickname DRAGON'S INFERNO and associated nose art, and was assigned to the 419th Test Squadron, 412nd Test Wing at Edwards AFB.

(Author)

(Author)

60-0050 is seen here on the ramp at Edwards AFB during October 1996. The red seeker cover of an AGM-142 can be seen under the Wing. The missile was loaded as part of a two aircraft test, in which the missile was launched from one B-52H and controlled by another B-52H. (Author)

(Brian C. Rogers)

60-0051 APPETITE FOR DESTRUCTION II

The aircraft has received CEM. On June 30, 1993, 60-0051 was assigned to the 92nd Bomb Wing at Fairchild AFB and had accumulated 16,197.1 flight hours. Next assigned to Minot AFB it was nicknamed PRAIRIE WARRIOR with associated nose art. By August 31, 1996 it was assigned to the 5th Bomb Wing and had accumulated 17,197.1 flight hours. As of August 1, 1998, 60-0051 was nicknamed APPETITE FOR DESTRUCTION II with associated nose art and assigned to the 23rd Bomb Squadron, 5th Bomb Wing at Minot AFB.

(Brian C. Rogers)

60-0051 is seen here at Minot AFB on April 1, 1998, with APPETITE FOR DESTRUCTION II nose art and the 23rd Bomb Squadron tail stripe. (Brian C. Rogers)

60-0051 is seen here at Minot AFB on November 15, 1995, with PRAIRIE WARRIOR nose art. A cruise missile pylon loaded with AGM-129 ACMs is under the Wing. (Brian C. Rogers)

60-0052

The aircraft has received CEM. On June 30, 1993, 60-0052 was assigned to the 92nd Bomb Wing at Fairchild AFB and had accumulated 16,816.5 flight hours. By August 31, 1996 it was assigned to the 2nd Bomb Wing and had accumulated 17,709.2 flight hours. As of August 1, 1998, 60-0052 was assigned to the 96th Bomb Squadron, 2nd Bomb Wing at Barksdale AFB.

ABOVE: 60-0052 is seen here at Barksdale AFB on June 30, 1997, with the red tail stripe of the 96th Bomb Squadron, 2nd Bomb Wing. (Author)
BELOW: 60-0052 is seen here in April 1993 with the 92nd Bomb Wing FC tail code and the blue tail cap of early Fairchild B-52Hs. (Ben Knowles)

60-0053

On June 30, 1993, 60-0053 was assigned to the 410th Bomb Wing at K.I. Sawyer AFB and had accumulated 12,337.8 flight hours. By August 31, 1996 it was assigned to the 2nd Bomb Wing and had accumulated 13,258.6 flight hours. As of August 1, 1998, 60-0053 was assigned to the 96th Bomb Squadron, 2nd Bomb Wing at Barksdale AFB.

60-0053 is seen here at Barksdale AFB on June 30, 1997, with the red tail stripe of the 96th Bomb Squadron, 2nd Bomb Wing. (Author)

(Author)

60-0054 MUD BUFF

On June 30, 1993, 60-0054 was assigned to the 410th Bomb Wing at K.I. Sawyer AFB and had accumulated 12,841.9 flight hours. By August 31, 1996 it was assigned to the 2nd Bomb Wing and had accumulated 13,577.9 flight hours. On August 8, 1995, shortly after takeoff, a massive inflight engine failure of the number 3 and 4 engines blew off both engines and the left inboard engine pylon. The aircraft continued to fly for over six hours until it was able to burn off enough fuel to safely land. As of August 1, 1998, 60-0054 was assigned to the 96th Bomb Squadron, 2nd Bomb Wing at Barksdale AFB. After a temporary repair, it was flown to Tinker AFB on August 6, 1998, for permanent repair including installation of new engines and an engine pylon. The landing gear and flaps were left down for the flight to Tinker AFB. 60-0054 is planned to be back in service in mid-1999.

60-0054 is seen here at Barksdale AFB on May 5, 1995, with the red tail stripe of the 96th Bomb Squadron, 2nd Bomb Wing. (Author)

60-0054 is seen here during June 1992 in the markings of the 644th Bomb Squadron, 410th Bomb Wing at K. I. Sawyer AFB. (David Brown)

60-0055 OLD CROW EXPRESS II
5th OPERATIONS GROUP FLAGSHIP

On June 30, 1993, 60-0055 was assigned to the 5th Bomb Wing at Minot AFB and had accumulated 13,540.5 flight hours. By August 31, 1996 it was assigned to the 2nd Bomb Wing and had accumulated 14,476.4 flight hours. It was marked as the 5th Operations Group (OG) Flagship with the nickname OLD CROW EXPRESS II and associated nose art, replacing 61-0031 which had transferred to Barksdale AFB. As of August 1, 1998, 60-0055 was assigned to the 23rd Bomb Squadron, 5th Bomb Wing at Minot AFB as an attrition reserve aircraft.

(Brian C. Rogers)

60-0055 is seen here at Minot AFB on August 27, 1996, with OLD CROW EXPRESS II nose art and marked as the 5th Operations Group (OG) Flagship. (Brian C. Rogers)

60-0055 is seen here on the ramp at Minot AFB on March 30, 1996, marked as the 5th Bomb Wing Flagship. (Brian C. Rogers)

(Brian C. Rogers)

60-0056 TAZ

On June 30, 1993, 60-0056 was assigned to the 410th Bomb Wing at K.I. Sawyer AFB and had accumulated 12,737.5 flight hours. By August 31, 1996 it was assigned to the 5th Bomb Wing and had accumulated 13,664.9 flight hours. As of August 1, 1998, 60-0056 was assigned to the 23rd Bomb Squadron, 5th Bomb Wing at Minot AFB as an attrition reserve aircraft.

60-0056 is seen here in 23rd Bomb Squadron markings. (Brian C. Rogers)

60-0056 is seen here at Minot AFB on February 17, 1996, with TAZ nose art and the tail stripe of the 5th Bomb Wing. (Brian C. Rogers)

60-0056 is seen here on July 22, 1993, in the markings of the 644th Bomb Squadron, 410th Bomb Wing at K. I. Sawyer AFB. (Norris Graser)

60-0057 NEMESIS

The aircraft has received the Conventional Enhancement Modification (CEM) and AGM-142 Have Nap capability. On June 30, 1993, 60-0057 was assigned to the 5th Bomb Wing at Minot AFB and had accumulated 14,016.2 flight hours. By August 31, 1996 it was assigned to the 2nd Bomb Wing and had accumulated 14,999.1 flight hours. As of August 1, 1998, 60-0057 was assigned to the 20th Bomb Squadron, 2nd Bomb Wing at Barksdale AFB.

(Brian C. Rogers)

60-0057 is seen here at Barksdale AFB on November 24, 1997, with NEMESIS nose art and the tail stripe of the 20th Bomb Squadron, 2nd Bomb Wing. (Brian C. Rogers)

60-0057 is seen here on takeoff from Ellsworth AFB during June 1994 with the tail stripe of the 5th Bomb Wing. (Author)

B-52H -160 -BW

60-0058

The aircraft has received CEM. On June 30, 1993, 60-0058 was assigned to the 92nd Bomb Wing at Fairchild AFB and had accumulated 13,700.1 flight hours. By August 31, 1996 it was assigned to the 2nd Bomb Wing and had accumulated 14,695.1 flight hours. As of August 1, 1998, 60-0058 was assigned to the 11th Bomb Squadron, 2nd Bomb Wing at Barksdale AFB.

BELOW: 60-0058 is seen here at Barksdale AFB on June 30, 1997, with the tail stripe of the 11th Bomb Squadron, 2nd Bomb Wing. (Author)

BELOW CENTER: 60-0058 is seen here at Minot AFB on June 23, 1995, with the tail stripe of the 5th Bomb Wing. (Brian C. Rogers)

60-0058 is seen here at Castle AFB on May 27, 1993, with the red and black checkered tail stripe of the 92nd Bomb Wing. (Author)

60-0059 LAISSEZ LE BON TEMPS ROULEZ
96th BOMB SQUADRON FLAGSHIP

The aircraft has received CEM. On June 30, 1993, 60-0059 was assigned to the 410th Bomb Wing at K.I. Sawyer AFB and had accumulated 12,383.9 flight hours. On August 1 thru 3, 1994 60-0059 accompanied 60-0008 on a 47 hour around the world flight. By August 31, 1996 it was assigned to the 2nd Bomb Wing and had accumulated 13,449.2 flight hours. As of August 1, 1998, 60-0059 was marked as the 96th Bomb Squadron Flagship and assigned to the 2nd Bomb Wing at Barksdale AFB.

As DUKE 02 the aircraft took part in the CALCM attack against Iraq on September 3, 1996, launching 7 of 8 CALCMs from the internal CSRL. The targets were command and control facilities and radar sites.

ABOVE RIGHT and RIGHT: (Author)

60-0053 is seen here at Barksdale AFB on April 29, 1994 (above) and on May 5, 1995 (below) marked as the 96th Bomb Squadron Flagship. (Brian C. Rogers above and Author below)

60-0060 IRON BUTTERFLY

The aircraft has received CEM. On June 30, 1993, 60-0060 was assigned to the 416th Bomb Wing at Griffiss AFB and had accumulated 13,092.7 flight hours. At Griffiss it was nicknamed OLD SMOKIE with associated nose art. After transferring to Minot, 60-0060 was given the nickname SCREMIN' GENIE with associated nose art. By August 31, 1996 it was assigned to the 5th Bomb Wing and had accumulated 14,083.6 flight hours. As of August 1, 1998, 60-0060 was marked as IRON BUTTERFLY with associated nose art, and was assigned to the 23rd Bomb Squadron, 5th Bomb Wing at Minot AFB.

(Tom Kaminski)

(Brian C. Rogers)

60-0060 is seen here at Nellis AFB on October 14, 1993, in the markings of the 668th Bomb Squadron, 416th Bomb Wing, Griffiss AFB, New York. (Brian C. Rogers)

60-0060 SCREMIN' GENIE. (Brian C. Rogers)

60-0060 IRON BUTTERFLY. (Brian C. Rogers)

60-0060 SCREAMIN GENIE is seen here at Nellis AFB on October 11, 1996, in the markings of the 23rd Bomb Squadron, 5th Bomb Wing, Minot AFB, North Dakota. (Brian C. Rogers)

60-0060 is seen here at Minot AFB on April 15, 1996, in the markings of the 23rd Bomb Squadron, 5th Bomb Wing carrying M 117(750 pound) bombs on a stub pylon and HSAB. (Brian C. Rogers)

60-0061 - 72nd BOMB SQUADRON FLAGSHIP

On June 30, 1993, 60-0061 was assigned to the 5th Bomb Wing at Minot AFB and had accumulated 11,622.0 flight hours. While assigned to Minot, 60-0061 was marked as the 72nd Bomb Squadron Flagship. By August 31, 1996 it was assigned to the 2nd Bomb Wing and had accumulated 12,427.0 flight hours. As of August 1, 1998, 60-0061 was assigned to the 11th Bomb Squadron, 2nd Bomb Wing at Barksdale AFB.

60-0061 is seen here at Minot AFB on August 10, 1995, marked as the 72nd Bomb Squadron Flagship. (Brian C. Rogers)

60-0061 is seen here at Barksdale AFB on October 15, 1997, in the markings of the 11th Bomb Squadron, 2nd Bomb Wing carrying M 117(750 pound) bombs on a stub pylon and HSAB. (Brian C. Rogers)

60-0062 CAJUN FEAR

The aircraft has received CEM and was one of the four B-52Hs to receive Rapid Eight AGM-142 Have Nap capability. It was also modified to the standard AGM-142 Have Nap Capability and in December 1992 it was the first B-52H to launch a live AGM-142. On June 30, 1993, 60-0062 was assigned to the 2nd Bomb Wing at Barksdale AFB and had accumulated 10,667.7 flight hours. By August 31, 1996 it was assigned to the 2nd Bomb Wing and had accumulated 11,638.6 flight hours. As of August 1, 1998, 60-0062 was named CAJUN FEAR with associated nose art and assigned to the 20th Bomb Squadron, 2nd Bomb Wing at Barksdale AFB.

(Brian C. Rogers)

60-0062 is seen here at Barksdale AFB on June 6, 1995, in the markings of the 20th Bomb Squadron, 2nd Bomb Wing carrying AGM-142 Have Nap on a stub pylon and HSAB. (Brian C. Rogers)

B-52H -165 -BW

61-0001

On June 30, 1993, 61-0001 was assigned to the 92nd Bomb Wing at Fairchild AFB. By August 31, 1996 it was assigned to the 2nd Bomb Wing and had accumulated 13,834.9 flight hours. As of August 1, 1998, 61-0001 was assigned to the 23rd Bomb Squadron, 5th Bomb Wing at Minot AFB as an attrition reserve aircraft.

BELOW: 61-0001 is seen here at Minot AFB on February 8, 1997, in the markings of the 23rd Bomb Squadron, 5th Bomb Wing. (Brian C. Rogers)

BELOW CENTER: 61-0001 is seen here taxiing at Minot AFB on March 17, 1997, in the markings of the 23rd Bomb Squadron, 5th Bomb Wing. (Brian C. Rogers)

61-0001 is seen here at Minot AFB on April 10, 1997, in the markings of the 72nd Bomb Squadron, 5th Bomb Wing. (Brian C. Rogers)

61-0002 EAGLES WRATH III
2nd OPERATIONS GROUP FLAGSHIP

The aircraft has received CEM. On June 30, 1993, 61-0002 was assigned to the 2nd Bomb Wing at Barksdale AFB and had accumulated 12,732.0 flight hours. While at Barksdale 61-0002 was given the nickname EAGLES WRATH III with associated nose art and marked as the 2nd Operations Group (OG) Flagship. By August 31, 1996 it was assigned to the 2nd Bomb Wing and had accumulated 13,810.5 flight hours. As of August 1, 1998, 61-0002 was assigned to the 20th Bomb Squadron, 2nd Bomb Wing at Barksdale AFB. It is still marked as the 2nd Operations Group (OG) Flagship and named EAGLES WRATH III, with the nose art updated as seen here.

ABOVE: (Author)

LEFT: Current nose art of EAGLES WRATH III. (Brian C. Rogers)

BELOW: 61-0002 is seen here at Barksdale AFB on May 5, 1995, THE EAGLE'S WRATH III marked as the 2nd Operations Group (OG) Flagship with the blue tail stripe of the 20th Bomb Squadron, 2nd Bomb Wing. (Author)

61-0002 is seen here at Barksdale on October 27, 1997, with modified nose art and marked as the 2nd OG Flagship. (Brian C. Rogers)

(Robert F. Dorr)

61-0003 BUFF BUNNY

The aircraft has received CEM. On June 30, 1993, 61-0003 was assigned to the 92nd Bomb Wing at Fairchild AFB and had accumulated 12,649.5 flight hours. As a result of the redesignation 92nd Bomb Wing to a Air Refueling Wing, 61-0003 was transferred to Minot. While at Minot it was assigned to the 23rd Bomb Squadron and given the nickname BUFF BUNNY with associated nose art. By August 31, 1996 it was assigned to the 2nd Bomb Wing and had accumulated 13,523.1 flight hours. As of August 1, 1998, 61-0003 was assigned to the 11th Bomb Squadron, 2nd Bomb Wing at Barksdale AFB.

61-0003 is seen here at Minot AFB on September 21, 1995, in the markings of the 23rd Bomb Squadron, 5th Bomb Wing. (Brian C. Rogers)

61-0003 is seen here at Barksdale AFB on June 30, 1997, with the tail stripe of the 11th Bomb Squadron, 2nd Bomb Wing. (Author)

61-0004 EL LOBO

The aircraft has received the Conventional Enhancement Modification (CEM) and was one of the four B-52Hs to receive Rapid Eight AGM-142 Have Nap capability. It was also modified to the standard AGM-142 Have Nap capability. On June 30, 1993, 61-0004 was assigned to the 410th Bomb Wing at K.I. Sawyer AFB and had accumulated 12,672.8 flight hours. By August 31, 1996 it was assigned to the 2nd Bomb Wing and had accumulated 13,684.0 flight hours. After transfer to Barksdale AFB it was nicknamed EL LOBO with associated nose art added. As of August 1, 1998, 61-0004 was assigned to the 20th Bomb Squadron, 2nd Bomb Wing at Barksdale AFB.

(Brian C. Rogers)

61-0004 is seen here at Barksdale AFB on November 24, 1997. LOBO is in the markings of the 20th Bomb Squadron, 2nd Bomb Wing. (Brian C. Rogers)

61-0005
5th OPERATIONS GROUP FLAGSHIP

The aircraft was not modified to carry CSRL. On June 30, 1993, 61-0005 was assigned to the 410th Bomb Wing at K.I. Sawyer AFB and had accumulated 14,336.9 flight hours. During 1995, while assigned to the 72nd Bomb Squadron, it was marked as the 5th Operations Group (OG) Flagship. By August 31, 1996 it was assigned to the 2nd Bomb Wing and had accumulated 15,075.4 flight hours. As of August 1, 1998, 61-0005 was assigned to the 23rd Bomb Squadron, 5th Bomb Wing at Minot AFB.

61-0005, the 5th OG Flagship, is seen here at Minot AFB on June 23, 1995, with the tail stripe of the 72nd Bomb Squadron, 5th Bomb Wing. (Brian C. Rogers)

61-0005 is seen here at Minot AFB on October 21, 1995, marked as the 5th Operations Group (OG) Flagship with the tail stripe of the 72nd Bomb Squadron, 5th Bomb Wing. (Brian C. Rogers)

61-0006

The aircraft has received CEM. On June 30, 1993, 61-0006 was assigned to the 5th Bomb Wing at Minot AFB and had accumulated 13,716.1 flight hours. By August 31, 1996 it was assigned to the 5th Bomb Wing and had accumulated 14,583.3 flight hours. As of August 1, 1998, 61-0006 was assigned to the 11th Bomb Squadron, 2nd Bomb Wing at Barksdale AFB. 61-0006 took part in Operation Desert Fox.

61-0006 is seen here at Minot AFB on July 18, 1995, with the tail stripe of the 23rd Bomb Squadron, 5th Bomb Wing. (Brian C. Rogers)

61-0006 is seen here at Barksdale AFB on May 8, 1996, with the tail stripe of the 11th Bomb Squadron, 2nd Bomb Wing as maintenance personnel replace the drag chute. (Author)

(Brian C. Rogers)

61-0007 GHOST RIDER

The aircraft has received CEM. On June 30, 1993, 61-0007 was assigned to the 5th Bomb Wing at Minot AFB and had accumulated 12,999.1 flight hours. By August 31, 1996 it was assigned to the 5th Bomb Wing and had accumulated 13,889.8 flight hours. It was given the nickname GHOST RIDER with associated nose art. As of August 1, 1998, 61-0007 was assigned to the 23rd Bomb Squadron, 5th Bomb Wing at Minot AFB. 61-0007 took part in Operation Desert Fox.

61-0007 GHOST RIDER is seen here at Minot AFB on April 14, 1996, with the tail stripe of the 23rd Bomb Squadron, 5th Bomb Wing. (Brian C. Rogers)

61-0007 is seen here taxiing at Minot AFB on May 13, 1996, with the tail stripe of the 23rd Bomb Squadron, 5th Bomb Wing. (Brian C. Rogers)

61-0008

The aircraft has received CEM. On June 30, 1993, 61-0008 was assigned to the 92nd Bomb Wing at Fairchild AFB and had accumulated 12,324.1 flight hours. By August 31, 1996 it was assigned to the 917th Wing and had accumulated 12,977.0 flight hours. As of August 1, 1998, 61-0008 was assigned to the 93rd Bomb Squadron, 917th Wing, AFRC at Barksdale AFB.

BELOW: 61-0008 is seen here on approach during August 1993 with the tail stripe and FC tail code of the 92nd Bomb Wing. (Ben Knowles)

BELOW CENTER: 61-0008 is seen here at Barksdale AFB on April 6, 1998, with the tail stripe of the 93rd Bomb Squadron, 917th Wing. The AFRES on the tail has been replaced with AFRC. (Brian C. Rogers)

61-0008 is seen here at Minot AFB on June 30, 1997, with the tail stripe of the 93rd Bomb Squadron, 917th Wing with AFRES on the tail. (Author)

61-0009

On June 30, 1993, 61-0009 was assigned to the 92nd Bomb Wing at Fairchild AFB. By August 31, 1996 it was assigned to the 2nd Bomb Wing and had accumulated 13,567.0 flight hours. As of August 1, 1998, 61-0009 was assigned to the 96th Bomb Squadron, 2nd Bomb Wing at Barksdale AFB.

61-0009 is seen here at Barksdale AFB on April 18, 1996, with the red tail stripe of the 96th Bomb Squadron, 2nd Bomb Wing. (Brian C. Rogers)

61-0010

The aircraft has received the Conventional Enhancement Modification (CEM) and AGM-142 Have Nap capability. On June 30, 1993, 61-0010 was assigned to the 5th Bomb Wing at Minot AFB and had accumulated 11,823.0 flight hours. By August 31, 1996 it was assigned to the 2nd Bomb Wing and had accumulated 13,741.8 flight hours. As of August 1, 1998, 61-0010 was assigned to the 20th Bomb Squadron, 2nd Bomb Wing at Barksdale AFB.

61-0010 is seen here at Barksdale AFB on May 18, 1995, carrying an AGM-142 Have Nap under the Wing, with the blue tail stripe of the 20th Bomb Squadron, 2nd Bomb Wing. (Brian C. Rogers)

61-0011 RAGIN' CAGIN
11th BOMB SQUADRON FLAGSHIP

The aircraft has received CEM. On June 30, 1993, 61-0011 was assigned to the 410th Bomb Wing at K.I. Sawyer AFB and had accumulated 11,892.1 flight hours. It is marked as the 11th Bomb Squadron Flagship, and nicknamed RAGIN' CAGIN with associated nose art. By August 31, 1996 it was assigned to the 2nd Bomb Wing and had accumulated 12,790.9 flight hours. As of August 1, 1998, 61-0011 was assigned to the 11th Bomb Squadron, 2nd Bomb Wing at Barksdale AFB.

(Brian C. Rogers)

(Brian C. Rogers)

61-0011, the 11th Bomb Squadron Flagship, is seen here at Barksdale AFB on August 7, 1994, (above) and on May 18, 1995 (below). (Both Brian C. Rogers)

On December 16, 1997, 61-0011 was setup at Barksdale AFB displaying the weapons capable of being carried by the B-52H. (Both Brian C. Rogers)

61-0012

The aircraft was not modified to carry CSRL. On June 30, 1993, 61-0012 was assigned to the 416th Bomb Wing at Griffiss AFB and had accumulated 13,805.8 flight hours. By August 31, 1996 it was assigned to the 2nd Bomb Wing and had accumulated 14,831.1 flight hours. 61-0012 transferred to Barksdale AFB and as of August 1, 1998 was assigned to the 11th Bomb Squadron, 2nd Bomb Wing at Barksdale AFB from the attrition reserve at Minot AFB.

(Tom Kaminski)

61-0012 seen taxiing back from a mission during July 1994. It is in the markings of the 416th Bomb Wing, Griffiss AFB, New York. (Tony Cassanova)

61-0012 is seen here at Barksdale AFB on May 18, 1995, with the tail stripe of the 11th Bomb Squadron, 2nd Bomb Wing carrying MK 84, 2000 pound bombs on a stub pylon-HSAB combination. (Brian C. Rogers)

61-0013

The aircraft was one of the four B-52Hs to receive Rapid Eight AGM-84 Harpoon modifications. On June 30, 1993, 61-0013 was assigned to the 5th Bomb Wing at Minot AFB and had accumulated 13,985.5 flight hours. By August 31, 1996 it was assigned to the 2nd Bomb Wing and had accumulated 15,350.9 flight hours. As of August 1, 1998, 61-0013 was assigned to the 96th Bomb Squadron, 2nd Bomb Wing at Barksdale AFB.

61-0013 is seen here on March 19, 1994, with the red and white candy stripe tail band of the 5th Bomb Wing. (Alan Winzerling)

61-0013 is seen here at Barksdale AFB on June 5, 1995, with the red tail stripe of the 96th Bomb Squadron, 2nd Bomb Wing. (Brian C. Rogers)

B-52H -170 -BW

61-0014

The aircraft was not modified to carry CSRL. On June 30, 1993, 61-0014 was assigned to the 410th Bomb Wing at K.I. Sawyer AFB and had accumulated 13,181.3 flight hours. By August 31, 1996 it was assigned to the 5th Bomb Wing and had accumulated 13,997.2 flight hours. As of August 1, 1998, 61-0014 was assigned to the 23rd Bomb Squadron, 5th Bomb Wing at Minot AFB as an attrition reserve aircraft.

61-0014 is seen here at Minot AFB on September 8, 1996, with the tail stripe of the 23rd Bomb Squadron, 2nd Bomb Wing. (Brian C. Rogers)

61-0014 is seen here at Barksdale AFB on May 18, 1995, with the tail stripe of the 11th Bomb Squadron, 2nd Bomb Wing. (Brian C. Rogers)

61-0015

The aircraft was not modified to carry CSRL. On June 30, 1993, 61-0015 was assigned to the 416th Bomb Wing at Griffiss AFB and had accumulated 14,123.3 flight hours. As a result of the inactivation of the 416th Bomb Wing, 61-0015 was transferred to the 2nd Bomb Wing at Barksdale AFB. By August 31, 1996 it was assigned to the 5th Bomb Wing and had accumulated 15,205.3 flight hours. As of August 1, 1998, 61-0015 was assigned to the 11th Bomb Squadron, 2nd Bomb Wing at Barksdale AFB from the attrition reserve at Minot AFB.

61-0015 is seen here taxiing at Minot AFB on March 28, 1996, with the tail stripe of the 11th Bomb Squadron, 2nd Bomb Wing and carrying MK 82 AIR 500 pound bombs. (Brian C. Rogers)

61-0015 is seen here at Barksdale AFB on May 18, 1995, with the blue tail stripe of the 20th Bomb Squadron, 2nd Bomb Wing. (Brian C. Rogers)

61-0016 FIRST STRIKE

The aircraft has received CEM. On June 30, 1993, 61-0016 was assigned to the 5th Bomb Wing at Minot AFB and had accumulated 13,375.4 flight hours. By August 31, 1996 it was assigned to the 2nd Bomb Wing and had accumulated 14,281.5 flight hours. As of August 1, 1998, 61-0016 was assigned to the 11th Bomb Squadron, 2nd Bomb Wing at Barksdale AFB.

(Brian C. Rogers)

61-0016 is seen here at Minot AFB on November 21, 1996, with the tail stripe of the 11th Bomb Squadron, 2nd Bomb Wing. (Brian C. Rogers)

(Author)

61-0017

On June 30, 1993, 61-0017 was assigned to the 92nd Bomb Wing at Fairchild AFB and had accumulated 13,355.4 flight hours. By August 31, 1996 it was assigned to the 917th Wing and had accumulated 14,069.8 flight hours. As of August 1, 1998, 61-0017 was assigned to the 93rd Bomb Squadron, 917th Wing at Barksdale AFB.

61-0017 is seen here at Ellsworth AFB on July 27, 1992, still wearing the 92nd Bomb Wing's SAC markings and the nose art which would soon become "politically incorrect." (Author)

61-0017 is seen wearing the blue tail stripe initially used by the 92nd Bomb Wing, after its transfer to ACC. (Ben Knowles)

61-0017 is seen here at Barksdale AFB on May 8, 1996, in the markings of the 93rd Bomb Squadron, 917th Wing, AFRES. (Author)

61-0018 DEADLY PRESCRIPTION

The aircraft was not modified to carry CSRL. On June 30, 1993, 61-0018 was assigned to the 410th Bomb Wing at K.I. Sawyer AFB and had accumulated 13,984.4 flight hours. As a result of the inactivation of the 410th Bomb Wing, 61-0018 was transferred to the 11th Bomb Squadron, 2nd Bomb Wing at Barksdale AFB. By August 31, 1996 it was assigned to the 5th Bomb Wing and had accumulated 15,043.8 flight hours. It was given the name DEADLY PRESC**RIP**TION with associated nose art, replacing 60-0020 which had transferred to Barksdale. As of August 1, 1998, 61-0018 was assigned to the 23rd Bomb Squadron, 5th Bomb Wing at Minot AFB as an attrition reserve aircraft.

(Brian C. Rogers)

61-0018, DEADLY PRESCRIPTION is seen here at Minot AFB on August 20, 1996, in the markings of the 23rd Bomb Squadron, 5th Bomb Wing. (Brian C. Rogers)

61-0018 is seen here at Barksdale AFB on May 5, 1995, in the markings of the 11th Bomb Squadron, 2nd Bomb Wing. (Author)

61-0019

The aircraft was one of the four B-52Hs to receive Rapid Eight AGM-84 Harpoon capability. On June 30, 1993, 61-0019 was assigned to the 410th Bomb Wing at K.I. Sawyer AFB and had accumulated 11,793.1 flight hours. On July 25, 1995 61-0019 was the first B-52H to launch a live AGM-84 Harpoon missile. By August 31, 1996 it was assigned to the 2nd Bomb Wing and had accumulated 12,852.3 flight hours. As of August 1, 1998, 61-0019 was assigned to the 96th Bomb Squadron, 2nd Bomb Wing at Barksdale AFB.

BELOW: 61-0019 is seen here at Barksdale AFB on May 5, 1995, in the markings of the 96th Bomb Squadron, 2nd Bomb Wing. (Author)

(Author)

61-0020

The aircraft has received CEM. On June 30, 1993, 61-0020 was assigned to the 416th Bomb Wing at Griffiss AFB and had accumulated 12,364 7 flight hours. It was given the name DEADLY PRESCRIPTION with associated nose art. By August 31, 1996 it had transferred to the 2nd Bomb Wing and had accumulated 13,326.7 flight hours. The nose art was removed. As of August 1, 1998, 61-0020 was assigned to the 96th Bomb Squadron, 2nd Bomb Wing at Barksdale AFB. 61-0020 took part in Operation Desert Fox.

61-0020 DEADLY PRESCRIPTION is seen here at Minot AFB on February 17, 1996, with the red and white candy stripe tail band of the 5th Bomb Wing. (Brian C. Rogers)

BELOW: 61-0020 seen at Griffiss AFB during July 1994 in the markings of the 416th Bomb Wing. (Jeff Wilson)

61-0021
The aircraft has received CEM. On June 30, 1993, 61-0021 was assigned to the 416th Bomb Wing at Griffiss AFB and had accumulated 11,902.4 flight hours. By August 31, 1996 it was assigned to the 917th Wing and had accumulated 12,887.4 flight hours. As of August 1, 1998, 61-0021 was assigned to the 93rd Bomb Squadron, 917th Bomb Wing, AFRC at Barksdale AFB.

61-0021 is seen here at Barksdale AFB on June 1, 1995, in the markings of the 93rd Bomb Squadron, 917th Wing, AFRES. (Brian C. Rogers)

61-0021 is seen at Griffiss AFB during April 1994 in the markings of the 668th Bomb Squadron, 416th Bomb Wing. (Tom Kaminski)

The external fuel tanks of the 93rd Bomb Squadron carry the Indian Head contained in the Unit Emblem. (Author)

61-0022

The aircraft has received CEM. On June 30, 1993, 61-0022 was assigned to the 5th Bomb Wing at Minot AFB and had accumulated 12,945.9 flight hours. By August 31, 1996 it was assigned to the 917th Wing and had accumulated 13,651.0 flight hours. As of August 1, 1998, 61-0022 was assigned to the 93rd Bomb Squadron, 917th Wing, AFRC at Barksdale AFB.

61-0022 is seen in these two photos at Barksdale on June 30, 1997, in the markings of the 93rd Bomb Squadron, 917th Wing, AFRC, at Barksdale AFB. (Both Author)

61-0023

The aircraft has received CEM and AGM-142 Have Nap capability. On June 30, 1993, 61-0023 was assigned to the 5th Bomb Wing at Minot AFB and had accumulated 8,531.2 flight hours. By August 31, 1996 it was assigned to the 2nd Bomb Wing and had accumulated 9,321.1 flight hours. As of August 1, 1998, 61-0023 was assigned to the 20th Bomb Squadron, 2nd Bomb Wing at Barksdale AFB.60-0023 took part in Operation Desert Fox.

61-0023 is seen here at Barksdale AFB on June 30, 1997, in the markings of the 20th Bomb Squadron, 2nd Bomb Wing. A lift trailer with a CSRL loaded with AGM-86 missile shapes can be seen behind the wing. (Author)

BELOW: 61-0023 is seen here at Barksdale AFB on October 15, 1997, in the markings of the 20th Bomb Squadron, 2nd Bomb Wing carrying an AGM-142 under the Wing. (Brian C. Rogers)

61-0024

The aircraft was one of the four B-52Hs to receive Rapid Eight AGM-84 Harpoon capability. On June 30, 1993, 61-0024 was assigned to the 410th Bomb Wing at K.I. Sawyer AFB and had accumulated 17,784.5 flight hours. By August 31, 1996 it was assigned to the 2nd Bomb Wing and had accumulated 18,734.8 flight hours. As of August 1, 1998, 61-0024 was assigned to the 96th Bomb Squadron, 2nd Bomb Wing at Barksdale AFB.

BELOW: 61-0024 is seen here at Barksdale AFB on May 18, 1995, in the markings of the 96th Bomb Squadron, 2nd Bomb Wing. (Brian C. Rogers)

61-0028

On June 30, 1993, 61-0028 was assigned to the 2nd Bomb Wing at Barksdale AFB and had accumulated 12,812.1 flight hours. By August 31, 1996 it was assigned to the 2nd Bomb Wing and had accumulated 13,517.9 flight hours. As of August 1, 1998, 61-0028 was assigned to the 96th Bomb Squadron, 2nd Bomb Wing at Barksdale AFB.

BELOW: 61-0028 is seen here at Barksdale AFB on June 30, 1997, in the markings of the 96th Bomb Squadron, 2nd Bomb Wing. (Author)

61-0029

The aircraft has received CEM. On June 30, 1993, 61-0029 was assigned to the 410th Bomb Wing at K.I. Sawyer AFB and had accumulated 12,790.1 flight hours. By August 31, 1996 it was assigned to the 917th Bomb Wing and had accumulated 13,558.5 flight hours. As of August 1, 1998, 61-0029 was assigned to the 93rd Bomb Squadron, 917th Wing, AFRC at Barksdale AFB.

61-0029 is seen here at Barksdale AFB on June 30, 1995, in the markings of the 93rd Bomb Squadron, 917th Wing, AFRES. (Author)

RIGHT: 61-0029 is seen here at Barksdale AFB on May 5, 1995, in the markings of the 93rd Bomb Squadron, 917th Wing, AFRES. In this photo, the engine nacelles have been removed for maintenance, exposing the engines. (Author)

61-0029 is seen here on August 8, 1992, in the markings of the 644th Bomb Squadron, 410th Bomb Wing K. I. Sawyer AFB, Michigan. (Norris Graser)

61-0030 Crashed on November 2, 1967 at Griffiss AFB, New York before change over to ACC

On November 2, 1967, 61-0030 was assigned to the 416th Bomb Wing at Griffiss AFB. It crashed on a go-around, killing six crew members; two crew members survived.

61-0031

The aircraft has received CEM. On June 30, 1993, 61-0031 was assigned to the 5th Bomb Wing at Minot AFB and had accumulated 11,853.5 flight hours. While at Minot the aircraft carried the nickname Old Crow Express II and its associated nose art. By August 31, 1996 it was assigned to the 2nd Bomb Wing and had accumulated 12,846.0 flight hours. It has been marked as the 2nd Bomb Wing Flagship. As of August 1, 1998, 61-0031 was assigned to the 11th Bomb Squadron, 2nd Bomb Wing at Barksdale AFB.

(Brian C. Rogers)

61-0031, OLD CROW EXPRESS II, is seen here at Minot AFB on June 11, 1995, with the red and white candy stripe tail band of the 5th Bomb Wing. (Brian C. Rogers)

61-0031, marked as the 2nd Bomb Wing Flagship, is seen here at Barksdale on October 15, 1997, after just having been loaded with MK 84 GP 2000 pound bombs on a stub pylon and HSAB. (Brian C. Rogers)

61-0032

On June 30, 1993, 61-0032 was assigned to the 2nd Bomb Wing at Barksdale AFB and had accumulated 12,175.0 flight hours. By August 31, 1996 it was assigned to the 917th Wing and had accumulated 13,015.9 flight hours. As of August 1, 1998, 61-0032 was assigned to the 93rd Bomb Squadron, 917th Wing, AFRC at Barksdale AFB.

61-0032 is seen here at Barksdale AFB on June 30, 1995, in the markings of the 93rd Bomb Squadron, 917th Wing, AFRES. (Author)

61-0033 Destroyed on November 14, 1975 in a ground fire at Minot AFB, North Dakota before change over to ACC

On November 14, 1975, 61-0033 was assigned to the 5th Bomb Wing at Minot AFB. It exploded and burned during ground refueling. Two ground crew members were killed.

61-0034 VIGILANCE

The aircraft was not modified to carry CSRL. On June 30, 1993, 61-0034 was assigned to the 410th Bomb Wing at K.I. Sawyer AFB and had accumulated 12,724.7 flight hours. By August 31, 1996 it was assigned to the 5th Bomb Wing and had accumulated 13,650.8 flight hours. The aircraft was nicknamed VIGILANCE with associated nose art. As of August 1, 1998, 61-0034 was assigned to the 23rd Bomb Squadron, 5th Bomb Wing at Minot AFB as an attrition reserve aircraft.

(Brian C. Rogers)

61-0034 is seen here at Minot AFB on March 28, 1996, in the markings of the 23rd Bomb Squadron, 5th Bomb Wing and is carrying MK 82 AIR 500 pound bombs on an AGM-28 pylon and MER beam combination. (Brian C. Rogers)

61-0034 is seen here at Minot AFB on September 21, 1995, with the red and white candy stripe tail band of the 5th Bomb Wing. (Brian C. Rogers)

61-0035

The aircraft was not modified to carry CSRL. On June 30, 1993, 61-0035 was assigned to the 410th Bomb Wing at K.I. Sawyer AFB and had accumulated 12,143.3 flight hours. Following the inactivation of the 410th Bomb Wing, 61-0035 was assigned to the 2nd Bomb Wing at Barksdale AFB. By August 31, 1996 it was assigned to the 5th Bomb Wing and had accumulated 13,017.8 flight hours. As of August 1, 1998, 61-0035 was assigned to the 23rd Bomb Squadron, 5th Bomb Wing at Minot AFB as an attrition reserve aircraft.

61-0035 is seen here on the alert ramp at Minot AFB on July 19, 1996, in the markings of the 23rd Bomb Squadron, 5th Bomb Wing. (Brian C. Rogers)

61-0035 is seen here at Barksdale AFB on May 5, 1995, in the markings of the 11th Bomb Squadron, 2nd Bomb Wing. (Author)

61-0036 BALTO

On June 30, 1993, 61-0036 was assigned to the 92nd Bomb Wing at Fairchild AFB and had accumulated 18,022.8 flight hours. By August 31, 1996 it was assigned to the 5th Bomb Wing and had accumulated 18,831.1 flight hours. 61-0036 was nicknamed BALTO with associated nose art and, as of August 1, 1998 was assigned to the 23rd Bomb Squadron, 5th Bomb Wing at Minot AFB as an attrition reserve aircraft.

(Brian C. Rogers)

61-0036, is seen here at Minot AFB on April 14, 1995, with the red and white candy stripe tail band of the 5th Bomb Wing. (Brian C. Rogers)

61-0036 is seen here at McConnell AFB during August 1993 in the markings of the 92nd Bomb Wing, Fairchild AFB, Washington. (Jerry Geer)

61-0036 is seen here at Minot AFB on June 10, 1996, in the markings of the 23rd Bomb Squadron, 5th Bomb Wing. (Brian C. Rogers)

61-0037 Crashed on January 21, 1969 near Minot AFB, North Dakota before change over to ACC

On January 21, 1969, 61-0037 was assigned to the 5th Bomb Wing at Minot AFB. It crashed after pitching up to a nose high attitude after lift-off. The aircraft stalled and crashed killing all six crew members.

61-0038

The aircraft has received the CEM. On June 30, 1993, 61-0038 was assigned to the 2nd Bomb Wing at Barksdale AFB and had accumulated 12,715.4 flight hours. By August 31, 1996 it was assigned to the 2nd Bomb Wing and had accumulated 13,672.9 flight hours. As of August 1, 1998, 61-0038 was assigned to the 11th Bomb Squadron, 2nd Bomb Wing at Barksdale AFB.

61-0038 is seen here at Barksdale AFB on May 5, 1995, in the markings of the 11th Bomb Squadron, 2nd Bomb Wing. (Author)

61-0039

The aircraft has received CEM. On June 30, 1993, 61-0039 was assigned to the 5th Bomb Wing at Minot AFB and had accumulated 12,293.0 flight hours. While at Minot 61-0039 was nicknamed APPETITE FOR DESTRUCTION. By August 31, 1996 it was assigned to the 2nd Bomb Wing and had accumulated 13,241.8 flight hours. As of August 1, 1998, 61-0039 was assigned to the 11th Bomb Squadron, 2nd Bomb Wing at Barksdale AFB.

61-0039, APPETITE FOR DESTRUCTION, is seen here at Minot AFB on July 10, 1995, in the markings of the 72nd Bomb Squadron, 5th Bomb Wing. (Brian C. Rogers)

(Brian C. Rogers)

61-0039, APPETITE FOR DESTRUCTION, is seen here at Minot AFB on April 10, 1995, with the red and white candy stripe tail band of the 5th Bomb Wing. (Brian C. Rogers)

61-0039 is seen here at Barksdale AFB on May 8, 1996 in the markings of the 11th Bomb Squadron, 2nd Bomb Wing. (Author)

61-0040

On June 30, 1993, 61-0040 was assigned to the 92nd Bomb Wing at Fairchild AFB. By August 31, 1996 it was assigned to the 5th Bomb Wing and had accumulated 18,643.5 flight hours. As of August 1, 1998, 61-0040 was assigned to the 23rd Bomb Squadron, 5th Bomb Wing at Minot AFB as an attrition reserve aircraft.

LEFT: 61-0040's tail cap was painted with experimental 23rd Bomb Squadron markings, reversing the colors of the standard tail stripe. The new color scheme was not adopted, and no other B-52s were painted with this stripe. (Brian C. Rogers)

61-0040 is seen here at Minot AFB on April 29, 1996, in the experimental markings of the 23rd Bomb Squadron, 5th Bomb Wing. (Brian C. Rogers)

61-0040 is seen here at Minot AFB on October 21, 1995, in the markings of the 72nd Bomb Squadron, 5th Bomb Wing. (Brian C. Rogers)

B-1B Lancer

The B-1 is a superior aircraft guaranteeing leading-edge bomber performance well past the turn of the century. Of all three bombers, the B-1 has the highest speed and greatest payload. After completion of an improvement program on its electronic countermeasures equipment called Defensive Systems Upgrade Program (DSUP) which is presently underway, the B-1 will be able to penetrate all but the highest-threat environments where only the B-2 can perform. Modern capabilities and sheer numbers make the B-1 the backbone of the conventional bomber force, and the only system that can compensate for the potential loss in target coverage as a result of the decision to finish the B-2 buy at only 21 aircraft. The B-1s low radar cross-section makes detection very difficult, and its ability to fly low and fast while carrying a larger payload, along with its advanced electronic countermeasures, enhances its survivability. The swing-wing design and turbofan engines provide greater range and high speed at low levels and also enhance the bomber's survivability. Wing sweep at the full-forward position allows a short takeoff roll and a fast base-escape profile for airfields under attack. Once airborne, the wings are positioned for maximum cruise distance or high-speed penetration. The B-1 now uses its new Global Positioning System (GPS), radar, and inertial navigation equipment enabling aircrews to globally navigate, update mission profiles and target coordinates in-flight, and precision bomb without the need for ground based navigation aids.

Destroying the bulk of defended time-critical targets early in a conflict by using direct attack or standoff munitions will be the B-1's primary role. Near-supersonic sustained airspeed and maneuverability give the B-1 the unique ability to fly with Air Force fighter aircraft in force packages much like the F-111's did in the Gulf; but instead of the four 2000 pound weapons of the F-111s, the B-1 can carry up to twenty-four.

By 2001, enhancements to the B-1 will complete its transition from a nuclear bomber to a more versatile bomber. It will receive more precise attack capabilities, both direct and standoff, to allow it to shoot and penetrate. These capabilities can work in tandem with the B-2's strong suit in leading the initial response, and will allow the theater commander to plan a broad range of air strike options even in the absence of substantial forces predeployed to the theater.

The DSUP is a high priority for the B-1. This ECM upgrade program is a good example of how the USAF is shifting away from Cold War missions. The original ECM specifications were for low altitude missions against Soviet air defenses. In regional conflicts, enemy air defense systems may be less than state-of-the-art but considerably more varied and unpredictable. Concentrations around valuable targets may be nearly as difficult as some of the programmed routes in the former Soviet Union. Implementing effective ECM and other capabilities is actually more crucial for conventional missions. The DSUP ECM program completes the aircraft ECM improvement that was stopped in fiscal year 1990 after restrictions enacted by Congress.

Originally the B-1 has only an unguided conventional bombing capability and a limited sea-mining potential. Certification for additional types of conventional munitions is underway. JDAM capability was added in January 1999. JSOW and JASSM will give the B-1 highly accurate weapons and expand the number of targets the B-1 can strike. The priority

funding for the JDAM also incorporates the MIL-STD-1760 interface and will replace the 1970s vintage 256K computer with high-speed, high-capacity computers fitted with the new software to employ smart weapons.

B-1 General Characteristics
Engines: Four General Electric F-101-GE-102 turbofan engine with afterburner
Thrust: 30,000-plus pounds with afterburner, per engine
Length: 146 feet (44.5 meters)
Wingspan: 137 feet (41.8 meters) extended forward, 79 feet (24.1 meters) swept aft
Height: 34 feet (10.4 meters)
Weight: Empty, approximately 190,000 pounds (86,183 kilograms)
Maximum Takeoff Weight: 477,000 pounds (214,650 kilograms)
Speed: 900-plus mph (Mach 1.2 at sea level)
Range: Intercontinental, unrefueled
Ceiling: Over 30,000 feet (9,000 meters)
Crew: Four (aircraft commander(pilot), copilot, offensive systems officer and defensive systems officer)
Armament: Up to 84 Mark 82 conventional 500-pound bombs and 30 CBU-87/89/97.
Also can be reconfigured to carry a wide range of nuclear weapons
Date Deployed: June 1985
Unit Cost: $200-plus million per aircraft

Inventory:
Active force, 72 (at four bases - Dyess AFB, Edwards AFB, Ellsworth AFB and Mountain Home AFB)
Air National Guard, 21 (at two bases McConnell AFB, Kansas Air National Guard and Robins AFB, Georgia Air National Guard)
Air Force Reserve Command, 0

B-1 OPERATIONAL UNITS

In June, 1992, the B-1s were assigned to four Bomb Wings, the 28th Bomb Wing at Ellsworth AFB, South Dakota; the 96th Bomb Wing at Dyess AFB, Texas; the 319th at Grand Forks AFB, North Dakota; and the 384th Bomb Wing at McConnell AFB, Kansas. As with the B-52s, as bomber units were inactivated, the active force of B-1 aircraft were grouped at two bases, Dyess AFB and Ellsworth AFB. Nine of the B-1s at Ellsworth AFB were assigned to the 34th Bomb Squadron, replacing the units B-52Gs which had been retired shortly before the 34th's April 4, 1994 move to Ellsworth AFB. As with the B-52Gs, the B-1s of the 34th Bomb Squadron were assigned to Ellsworth as a geographically separated unit (GSU) of the 366th Wing Based at Mountain Home AFB, Idaho. The B-1 support facilities of the 384th Bomb Group at McConnell AFB, Kansas were transferred to the 184th Bomb Group of the Kansas Air National Guard with the guard unit picking up 11 B-1s during July 1994. The B-1s of the 319th Bomb Group left Grand Forks AFB during early 1994. The Georgia Air National Guard became the second guard unit to fly the B-1 when the 116th gave up its F-15s and moved from Dobbins ARB in Marietta, Georgia, to its new home 125 miles south at Robins AFB in Warner Robins, Georgia. On April 1, 1996, the unit was officially redesignated as the 116th Bomb Wing equipped with the B-1B Lancer.

A B-1 is seen taking of as part of an exercise with the 34th Bomb Squadron deployed to Elmendorf AFB, Alaska, as part of Operation Amalgam Warrior during April 1996. (Dave Pasquini)

DYESS AFB, TEXAS

96th BOMB WING

The 96th Bomb Wing traces its lineage back to the 96 Bomb Group, activated on July 15, 1942, and was assigned to European Theater of Operations and North Africa during World War II. The 96th Bomb Group flew B-17s. The Group was inactivated on November 29, 1945.

On November 18, 1953, the 96th Bombardment Wing (Heavy) was activated at Altus AFB, Oklahoma, equipped with the B-47 Stratojet. On September 8, 1957, the Bomb Wing moved to Dyess AFB, Texas. During September of 1963 the Bomb Wing replaced the B-47 with the B-52E. In March 1970 the Bomb Wing replaced the B-52E with the B-52D. In October 1982 the B-52D was replaced by the B-52H. In 1982 the Bomb Wing converted to ALCM equipped B-52Gs. The last B-52G left Dyess on January 18, 1985.

96th Bomb Wing

The 96th Bombardment Wing transitioned to B-1 with the official arrival of 82-0001 standing in for 83-0065 STAR OF ABILENE (which was grounded at Offutt AFB with a broken air cooling duct) on June 27 1985. The 96th Bomb Wing completed its transition to the B-1 with the arrival of 85-0072 on February 27, 1987. In September of 1991 the 96th Bombardment Wing (Heavy) was redesignated as the 96th Wing and was assigned to the newly created ACC on June 1, 1992. The 337th Bombardment Squadron, the 338th Combat Crew Training Squadron, and 917th Air Refueling Squadron were assigned to the 96th Wing when it transferred to ACC. Aircraft of the 96th Bomb Wing carried DY (**DY**ess AFB) tail codes and most have nose art. Different than other bomber units, some of the Dyess AFB B-1s have the nose art on both sides of the aircraft. The Wing's motto is E Sempre L Ora' (It's Always The Hour).

337th Bomb Squadron

The 337 Bomb Squadron was formed out of assets of the 39th Bomb Squadron, moved to Dyess AFB from Walker AFB on September 15, 1963, and was equipped with the B-52E until March 1970 and the B-52D until October 1, 1982. The 337th began conversion to the B-52H with the arrival of the first aircraft on May 3, 1982, putting the first B-52H on alert on October 1, 1982. The 337th began phasing down in the fall of 1984, transferring its last B-52H to Carswell AFB on January 22, 1985, in preparation for the arrival of the first B-1. The B-1s started arriving in June of 1985. The 337th Bomb Squadron was the Wing's operational squadron until it became the Formal Training Unit (FTU) replacing the 338th CCTS on October 1, 1993. The 9th Bomb Squadron was activated on the same day becoming the Wing's operational squadron. The 337th was inactivated on October 1,1994, after being replaced as the FTU by the 28th Bomb Squadron.

The 337th aircraft wore DY tail codes and a red tail stripe with a yellow-gold dragon until becoming the training squadron. At that time they inherited the blue and white checked tail stripes of the 338th CCTS.

(Craig Kaston)

338th Combat Crew Training Squadron

The 338th Combat Crew Training Squadron (CCTS) was activated at Dyess AFB during January 1985 as the 4018 CCTS with the assigned mission of training B-1 aircrews for assignment to the SAC B-1 Bomb Wings. The 4018 CCTS was redesignated the 338th SBTS (Strategic Bombardment Training Squadron) and later designated the 338th CCTS. The 338th CCTS was inactivated on October 1, 1993, with

the 337th taking over the FTU mission from the 338th. The 338th CCTS wore DY tail codes and blue and white checked tail stripe.

(Brian C. Rogers)

7th BOMB WING

On October 1, 1993, the 7th Wing moved from Carswell AFB, Fort Worth, Texas to Dyess AFB, Texas, flying both the B-1 Lancer and the C-130 Hercules. This unique structure of bombing and airlift under one Wing remained intact until April 1, 1997, when the Air Force transferred all C-130s to Air Mobility Command. That same day, the 317th Airlift Group stood

7th Bomb Wing

(Author)

up at Dyess, encompassing all Dyess C-130 assets and the 7th Wing became the 7th Bomb Wing. The 7th Wing aircraft carried the same DY tail code as the 96th Wing and kept the nose art, some on both sides of the aircraft. The 7th Wing's motto is Mors Ab Alto (Death From On High).

9th Bomb Squadron

The 9th Bomb Squadron was activated at Dyess AFB on October 1, 1993, taking over the mission of the 337th Bomb Squadron which became the Training Squadron. The 9th Bomb Squadron wear DY tail codes and a black tail stripe with a white bat centered in the stripe.

(Author)

28th Bomb Squadron

The 28th Bomb Squadron moved without personnel or equipment from the 384th Bomb Group at McConnell AFB on October 1, 1994. It took over the air crew training responsibility from the 337th CCTS becoming the B-1 FTU. The 337th Bomb Squadron inactivated on October 1, 1994. The 28th Bomb Squadron wears the DY tail code and the blue and white checkered tail stripe it inherited from the 337th/338th Squadrons.

(Author)

ELLSWORTH AFB, SOUTH DAKOTA

28th BOMB WING

In 1986 the Ellsworth AFB and the 28th Bomb Wing made extensive preparations to transfer its B-52s and become the new home for the new B-1 Lancer. The last of the 28th Bomb Wing B-52H's left in early 1986. The Wing received the first of 35 B-1 bombers starting with the official arrival of the first B-1 on January 21, 1987. The 28th Bomb Wing completed its transition to the B-1 with the arrival of its last B-1 on September 15, 1987. The 28th Bomb Wing was initially made up of two bomb squadrons, the 37th Bombardment Squadron and the 77th Bombardment Squadron and one air refueling squadron, the 28th Air Refueling Squadron. In September 1991 the 28th Bombardment Wing (Heavy) was redesignated as the 28th Wing and was assigned to the newly created Air Combat Command on June 1, 1992. Upon assignment to ACC, the Wing was redesignated as the 28th Bomb Wing. The 28th Air Refueling Squadron was transferred from the 28th Wing to the Air Mobility Command (AMC) on June 1, 1992.

The inactivation of one of Ellsworth's oldest units, the 77th Bomb Squadron occurred in 1995. While the unit was inactivated to save Air Force dollars for development of new follow-on B-1 munitions, the organizations aircraft remained at Ellsworth (in a flying reserve status) with the 37th Bomb Squadron.

(Author)

Ellsworth's B-1s carry an EL tail code (**EL**lsworth AFB), and most have nose art on the left side of the forward fuselage. The Wing's motto is Guardian Of The North.

37th Bombardment Squadron

The 37th Bomb Squadron was activated at Ellsworth on July 1, 1977, as the new second Bomb Squadron of the 28th Bomb Wing. The 28th Bomb Wing was at that time transitioning from a Bomb Wing with a single squadron of B-52Gs to a two squadron B-52H Bomb Wing. On October 1, 1982 the 37th was inactivated. In October 1985, the 37th was reactivated and became the second 28th Bomb Wing B-1 squadron. The squadron currently flies the B-1. The 37th Bomb Squadron carries EL tail codes with a black stripe edged in gold and TIGERS also in gold.

(Author)

77th Bombardment Squadron

The 77th Bomb Squadron has been resident with the 28th Bomb Wing at Ellsworth since the 28th arrived at Ellsworth in 1947. The 77th converted from the B-36 to the B-52D in June 1957, and from the B-52D to B-52G during 1971. On May 2, 1977, transition to the B-52H began, with conversion complete on July 20, 1977. In October 1985, the 77th began phasing down B-52H operations for conversion to the B-1. The squadron flew the B-1 until it was inactivated on March 31, 1995. Prior to inactivation the squadron carried EL tail codes with a blue stripe containing a silhouette of Mt. Rushmore.

Original 77th Bomb Squadron patch

(Author)

It was again activated at Ellsworth AFB on April 1997 as the geographically separated 34th Bomb Squadron completed its transfer to its home at the 366th Wing, Mountain Home AFB, Idaho. By June 1998 the 77th had six of its B-1s, which came out of the reconstitution reserve and balanced those lost by the departure of the 34th BS. The 77th wears EL tail codes and a maroon stripe edged in yellow, with an eagle's head and the words WAR EAGLES in yellow within the stripe.

(Author)

Present 77th Bomb Squadron "War Eagles" patch

366th Wing, 34th Bombardment Squadron

In March, 1994 Ellsworth AFB added the 34th Bomb Squadron, a geographically separated unit awaiting airfield upgrades before it could return to its parent organization, the 366 Wing, at Mountain Home AFB, Idaho. The 34th's B-1s are part of the 366th Wing, which also includes F-15C/D Eagles, F-15E Strike Eagles, F-16 Fighting Falcons, and KC-135Rs. The 34th Bomb Squadron and its B-1s moved to Mountain Home AFB between August, 1996 and April 1, 1997. While at Ellsworth, the B-1s of the 34th Bomb Squadron carried MO (**MO**untian Home) tail codes, and a tail stripe of a red thunderbird followed by red lines over a black stripe. None of the 34th Bomb Squadron's B-1s have nose art.

57th WING, DETACHMENT 4, B-1 DIVISION

On October 1, 1989 the Strategic Weapons School (SWS) was activated at Ellsworth AFB. After the 99th Strategic Weapons Wing (SWW) was activated on August 10, 1989, the SWS was assigned to the 99th SWW. The SWS initially conducted B-52 Weapons Instructor training. During the 1991 -1992 time frame a B-1 Division was formed. SAC was planning addition of a tanker course, and in September 1991 the SWS was renamed the Bomber-Tanker Employment School. In June 1992, following the inactivation of SAC, the school was inac-

tivated and replaced by Detachment 1 of the USAF Weapons School, which included both B-52 and B-1 divisions. The B-52 Division moved to Barksdale AFB during 1995. As the 99th Wing was inactivating, the two bomber divisions picked up the test functions previously accomplished by the 99th Wing. As a result both Divisions, now with both Weapons School and Test functions, became part of the 57th Wing at Nellis AFB. The B-1 Division at Ellsworth became Det 4, 57th Wing, and the B-52 Division at Barksdale became Det 5, 57th Wing. The B-1 Weapons School does not have any aircraft of own and borrows 28th Bomb Wing aircraft when necessary.

53rd Wing, 79th Test and Evaluation Group, Detachment 2 B-1 TEST

B-1 operational test and evaluation is accomplished by Detachment 2 at Ellsworth AFB. Detachment 2 is part of the 79th Test and Evaluation Group (TEG), 53rd Wing which is based at Eglin AFB, Florida. Transfer of the test functions from Det 4 occurred when the Air Warfare Center restructure took place in October 1996. As part of the restructure, the B-1 test functions were transferred from the 57th Wing Detachment to Det 2 of the 79th TEG. The B-1 TEST detachment does not have any aircraft of its own and borrows other unit's aircraft when necessary.

GRAND FORKS AFB, NORTH DAKOTA

319th BOMB GROUP

On February 1, 1963, the 319 Bombardment Wing (Heavy) was activated at Grand Forks AFB replacing the 4133rd Strategic Wing. The bomb wing had two operational squadrons, the 46th Bombardment Squadron equipped with the B-52H Stratofortress and the 905th Air Refueling Squadron equipped with the KC-135A Stratotanker. In 1982 the bomb wing converted to ALCM equipped B-52Gs. The 319th was operational ready with ALCMs in 1984 and, in late 1984 through the first half of 1985 participated in a series of test launches of the ALCM over the U.S. and Canada.

The 319th Bombardment Wing (Heavy) transitioned to B-1 and KC-135R aircraft with the official arrival of both B-1 and KC-135R aircraft on October 16, 1987. The 319th Bomb Wing completed its transition to the B-1 with the arrival of its last B-1 on January 12, 1988. The 46th Bombardment Squadron and the 905th Air Refueling Squadron were assigned to the 319th Bomb Wing. In September of 1991 the 319th Bombardment Wing (Heavy) was redesignated as the 319th Wing and was assigned to the newly created Air Combat Command on June 1, 1992. Upon assignment to ACC, the wing was redesignated as the 319th Bomb Wing. The 319th Bomb Wing was inactivated and the 319th Bomb Group was activated on October 1, 1993. The 46th Bomb Squadron was assigned to the newly created 319th Bomb Group. The KC-135s were assigned to the newly activated 319th Air Refueling Wing (ARW) part of the Air Mobility Command (AMC). The B-1s were reassigned to other bases during early 1994, and the Bomb Group was inactivated on July 16, 1994.

The 319th's B-1s wore GF (**G**rand **F**orks) tail codes and a blue and orange tail stripe with an emblem called a "Sunflake", half sunburst and half snowflake. Most of Grand Fork's B-1s retained the nose art applied during the late 1980s. The Wing's motto is Defensores Libertatis (Defenders of Liberty).

(Brian C. Rogers)

46th Bombardment Squadron

The 46th Bomb Squadron was activated at Grand Forks AFB on February 1, 1963, replacing the 30th Bomb Squadron of the 4133rd Strategic Wing. From 1980 the 46th participated in the Strategic Projection Force with the 23rd at Minot AFB. During the summer of 1982 the 46th converted to the ALCM capable B-52G. The 46th transferred their last B-52G on December 4, 1986 in preparation for transition to the B-1. The squadron flew the B-1 as part of the 319th Bomb Wing and later the 319th Bomb Group from October, 1987 until the unit was inactivated on July 16, 1994.

McCONNELL AFB, KANSAS

384th BOMB GROUP

On December 1, 1972, the 384th Air Refueling Wing (Heavy) was activated at McConnell AFB, Kansas. On July 2, 1984, the 384th AREFW became the first wing to be equipped with the KC-135R re-engined Stratotanker. By September, 1985, 27 of the re-engined tankers had been delivered to the 384th AREFW from Boeing's Wichita facility across the runway from McConnell.

384th Bomb Group

On July 1, 1987, the wing was redesignated the 384th Bombardment Wing (Heavy) in preparation for arrival of the B-1. The 384th officially received its first B-1 on February 29, 1988. In September of 1991 the 384th Bombardment Wing (Heavy) was redesignated as the 384th Wing and was assigned to the newly created Air Combat Command on June 1, 1992. Upon assignment to ACC, the wing was redesignated as the 384th Bomb Wing. The 384th Bomb Wing was redesignated the 384th Bomb Group on January 1, 1994, retaining the 28th Bomb Squadron. The KC-135s were assigned to the newly activated 22nd Air Refueling Wing (ARW) part of the Air Mobility Command (AMC). The B-1s were reassigned to other units, including the 184th Bomb Group of Kansas Air National Guard during the first half of 1994, and the 384th Bomb Group was inactivated on October 1, 1994.

The 384th's B-1s carried OZ tail codes (related to the Kansas connection to the "Land of **OZ**"). Their tail stripe was a blue stripe with an image of the "Keeper Of The Plains" statue, and the words "KEEPER OF THE PLAINS." Very few of the 384th B-1s had nose art. The Wing's motto is Keep The Show On The Road.

28th Bomb Squadron

At Homestead AFB, the 28th Bomb Squadron was one of the first original six B-52H units, receiving their first B-52H on February 7, 1962. During June and July, 1968, the 28th transferred its B-52Hs to the 17th Bomb Wing at Wright-Patterson AFB phasing down operations at Homestead AFB. On July 25, 1968, the 28th moved without personnel or aircraft to Robins AFB Georgia replacing the 781st Bomb Squadron of the 465 Bomb Wing. At Robins the 28th flew the B-52G. On October 1, 1983, the 28th was inactivated in conjunction with the 19th Bomb Wing being redesignated the 19th Air Refueling Wing. The 28th was reactivated at McConnell AFB in July, 1987 in preparation for the arrival of the B-1. The 28th continued to fly the B-1 with both the 384th Bomb Wing and the 384th Bomb Group. The 28th Bomb Squadron moved without personnel and equipment to Dyess AFB on October 1, 1994, replacing the 338th Bomb Squadron and taking over B-1 air crew training responsibilities.

(Author)

(Author)

KANSAS AIR NATIONAL GUARD 184th BOMB WING

The 184th Bomb Wing has a proud history and a wealth of tradition. The unit was activated to federal service during World War II, the Korean Conflict, and the Viet Nam Conflict. In January 1987, the 184th was tasked to conduct conversion and upgrade training in the F-16. In July 1994, the 184th Fighter Group was designated at the 184th Bomb Group and again became part of the Air Combat Command, flying the B-1 Lancer. The 184th Bomb Group was redesignated the 184th Bomb Wing on October 1, 1995. The 184th was the first Air National Guard unit to fly bombers.

The 184th Bomb Wing Jayhawk's B-1s do not carry tail codes. The tail stripe is a red stripe with KANSAS in white. The Air National Guard insignia is modified with the stylized aircraft being replaced by B-1 silhouettes. Nose art is being added to the Kansas ANG B-1s.

(Author)

127th Bomb Squadron

With the arrival of the B-1s, the 127th Fighter Squadron was redesignated the 127th Bomb Squadron and began to train with its new aircraft. 127th continues to fly its 11 B-1s.

MOUNTAIN HOME AFB, IDAHO

366th WING

The 366th Wing, USAF's Air Expeditionary Wing is based at Mountain Home AFB, Idaho. In addition to the B-1s of the 34th Bomb Squadron, the wing operates F-15C/Ds assigned to 390th Fighter Squadron, F-15Es assigned to the 391st Fighter Squadron, F-16C/Ds assigned to the 389th Fighter Squadron, and KC-135R tankers assigned to the 22nd Air Refueling Squadron. All 366th aircraft wear MO (**MO**untain Home) tail codes.

(Dave Pasquini)

366th Wing

34th Bomb Squadron

On April 1, 1994, the 34th Bomb Squadron moved to Ellsworth AFB, South Dakota from Castle AFB, California. At the same time the Squadron's B-52Gs were retired, making way for the squadron to be equipped with the B-1 Lancer. The B-1s operated from Ellsworth AFB, while assigned as a geographically separate unit (GSU) of the 366th. A gradual transfer of the B-1s from Ellsworth to Mountain Home began in August, 1996. The Squadron completed a move to Mountain Home on April 1, 1997, when its flag was officially transferred to the 366th Gunfighter's home base. The B-1s of the 34th Bomb Squadron carried MO tail codes and a tail stripe of a red thunderbird followed by red lines over a black stripe. None of the 34th Bomb Squadron's B-1s carry nose art.

Four B-1s of the 34th Bomb Squadron are seen here on the ramp at Elmendorf AFB, Alaska during Operation Amalgam Warrior in April 1996. (Dave Pasquini)

ROBINS AFB, GEORGIA

GEORGIA AIR NATIONAL GUARD 116th BOMB WING

The 116th Bomb Wing Georgia Air National Guard is currently assigned to Robins AFB, Georgia, and equipped with the B1-B Lancer. The 116th BW has both federal and state missions. Upon mobilization for national emergency or war, the wing would augment regular forces and possibly deploy to an overseas location. Throughout its history, the 116th has been involved in every major conflict including World War II, Korea, Vietnam, and Desert Storm. In 1986 it began operating McDonnell Douglas F-15A/F-15B Eagle fighters.

For almost 50 years, Dobbins ARB in Marietta, Georgia, was home to the 116th but, in 1996, the unit was relocated 125 miles south to Robins AFB in Warner Robins, Georgia. On April 1, 1996, the unit was officially redesignated as the 116th Bomb Wing and began its new mission with the B1-B Lancer.

The B-1s of the Georgia Air National Guard wear GA (**G**eorgi**A**) tail codes. They have not yet added a unit tail stripe, and are in the process of adding nose art.

(Nate Leong)

128th Bomb Squadron

The 128th Fighter Squadron replaced its F-15s with the B-1 and was redesignated as the 128th Bomb Squadron on April 1, 1996.

ACC B-1 AIRCRAFT ASSIGNMENTS 1992-1998

SERIAL NUMBER	JUNE 1, 1992 ASSIGNMENT	AUGUST 31, 1996 ASSIGNMENT	AUGUST 1, 1998 ASSIGNMENT
82-0001	28th BW Ellsworth AFB	Disassembled May 1995	
83-0065	96th BW Dyess AFB	7th BW Dyess AFB	9th BS, 7th BW Dyess AFB
83-0066	96th BW Dyess AFB	7th BW Dyess AFB	28th BS, 7th BW Dyess AFB
83-0067	96th BW Dyess AFB	7th BW Dyess AFB	9th BS, 7th BW Dyess AFB
83-0068	96th BW Dyess AFB	7th BW Dyess AFB	28th BS, 7th BW Dyess AFB
83-0069	96th BW Dyess AFB	7th BW Dyess AFB	28th BS, 7th BW Dyess AFB
83-0070	96th BW Dyess AFB	7th BW Dyess AFB	28th BS, 7th BW Dyess AFB
83-0071	96th BW Dyess AFB	7th BW Dyess AFB	26th BS, 7th BW Dyess AFB
84-0049	412th TW Edwards AFB	412th TW Edwards AFB	419th TES, 412 TW Edwards AFB
84-0050	96th BW Dyess AFB	7th BW Dyess AFB	28th BS, 7th BW Dyess AFB
84-0051	96th BW Dyess AFB	7th BW Dyess AFB	9th BS, 7th BW Dyess AFB
84-0052	Crashed September 28, 1987		
84-0053	96th BW Dyess AFB	7th BW Dyess AFB	28th BS, 7th BW Dyess AFB
84-0054	96th BW Dyess AFB	7th BW Dyess AFB	28th BS, 7th BW Dyess AFB
84-0055	96th BW Dyess AFB	7th BW Dyess AFB	28th BS, 7th BW Dyess AFB
84-0056	96th BW Dyess AFB	7th BW Dyess AFB	28th BS, 7th BW Dyess AFB
84-0057	96th BW Dyess AFB	7th BW Dyess AFB	Crashed February 18, 1998 while assigned to 9th BS, 7th BW Dyess AFB
84-0058	96th BW Dyess AFB	7th BW Dyess AFB	9th BS, 7th BW Dyess AFB
85-0059	96th BW Dyess AFB	116th BW GA ANG	128th BS, 116th BW GA ANG
85-0060	28th BW Ellsworth AFB	184th BW KS ANG	127th BS, 184th BW KS ANG
85-0061	28th BW Ellsworth AFB	28th BW Ellsworth AFB	128th BS, 116th BW GA ANG
85-0062	96th BW Dyess AFB	7th BW Dyess AFB	9th BS, 7th BW Dyess AFB
85-0063	Crashed November 8, 1988		
85-0064	28th BW Ellsworth AFB	184th BW KS ANG	128th BS, 184th BW KS ANG
85-0065	96th BW Dyess AFB	7th BW Dyess AFB	128th BS, 116th BW GA ANG
85-0066	28th BW Ellsworth AFB	28th BW Ellsworth AFB	37th BS, 28th BW Ellsworth AFB
85-0067	96th BW Dyess AFB	7th BW Dyess AFB	28th BS, 7th BW Dyess AFB
85-0068	412th TW Edwards AFB	412th TW Edwards AFB	419th TES, 412th TW Edwards AFB
85-0069	96th BW Dyess AFB	184th BW KS ANG	127th BS, 184th BW KS ANG
85-0070	96th BW Dyess AFB	184th BW KS ANG	127th BS, 184th BW KS ANG
85-0071	96th BW Dyess AFB	7th BW Dyess AFB	128th BS, 116th BW GA ANG
85-0072	96th BW Dyess AFB	7th BW Dyess AFB	9th BS, 7th BW Dyess AFB
85-0073	96th BW Dyess AFB	7th BW Dyess AFB	28th BS, 7th BW Dyess AFB
85-0074	96th BW Dyess AFB	7th BW Dyess AFB	9th BS, 7th BW Dyess AFB
85-0075	28th BW Ellsworth AFB	28th BW Ellsworth AFB	37th BS, 28th BW Ellsworth AFB
85-0076	Crashed November 17, 1988		
85-0077	28th BW Ellsworth AFB	28th BW Ellsworth AFB	77th BS, 28th BW Ellsworth AFB
85-0078	28th BW Ellsworth AFB	28th BW Ellsworth AFB	Crashed September 19, 1997 while assigned to 37th BS, 28th BW Ellsworth AFB
85-0079	28th BW Ellsworth AFB	28th BW Ellsworth AFB	37th BS, 28th BW Ellsworth AFB
85-0080	384th BW McConnell AFB	184th BW KS ANG	127th BS, 184th BW KS ANG
85-0081	384th BW McConnell AFB	184th BW KS ANG	127th BS, 184th BW KS ANG
85-0082	96th BW Dyess AFB	7th BW Dyess AFB	9th BS, 7th BW Dyess AFB
85-0083	28th BW Ellsworth AFB	28th BW Ellsworth AFB	37th BS, 28th BW Ellsworth AFB
85-0084	28th BW Ellsworth AFB	28th BW Ellsworth AFB	37th BS, 28th BW Ellsworth AFB
85-0085	28th BW Ellsworth AFB	28th BW Ellsworth AFB	77th BS, 28th BW Ellsworth AFB
85-0086	28th BW Ellsworth AFB	28th BW Ellsworth AFB	37th BS, 28th BW Ellsworth AFB

85-0087	28th BW Ellsworth AFB	28th BW Ellsworth AFB	77th BS, 28th BW Ellsworth AFB
85-0088	28th BW Ellsworth AFB	184th BW KS ANG	127th BS, 184th BW KS ANG
85-0089	28th BW Ellsworth AFB	28th BW Ellsworth AFB	128th BS, 184th BW KS ANG
85-0090	28th BW Ellsworth AFB	28th BW Ellsworth AFB	37th BS, 28th BW Ellsworth AFB
85-0091	28th BW Ellsworth AFB	28th BW Ellsworth AFB	37th BS, 28th BW Ellsworth AFB
85-0092	28th BW Ellsworth AFB	28th BW Ellsworth AFB	128th BS, 116th BW GA ANG
86-0093	28th BW Ellsworth AFB	28th BW Ellsworth AFB	37th BS, 28th BW Ellsworth AFB
86-0094	28th BW Ellsworth AFB	28th BW Ellsworth AFB	77th BS, 28th BW Ellsworth AFB
86-0095	96th BW Dyess AFB	184th BW KS ANG	127th BS, 184th BW KS ANG
86-0096	28th BW Ellsworth AFB	28th BW Ellsworth AFB	37th BS, 28th BW Ellsworth AFB
86-0097	319th BW Grand Forks AFB	28th BW Ellsworth AFB	77th BS, 28th BW Ellsworth AFB
86-0098	28th BW Ellsworth AFB	28th BW Ellsworth AFB	128th BS, 116th BW GA ANG
86-0099	28th BW Ellsworth AFB	28th BW Ellsworth AFB	37th BS, 28th BW Ellsworth AFB
86-0100	96th BW Dyess AFB	7th BW Dyess AFB	28th BS, 7th BW Dyess AFB
86-0101	384th BW McConnell AFB	7th BW Dyess AFB	28th BS, 7th BW Dyess AFB
86-0102	28th BW Ellsworth AFB	28th BW Ellsworth AFB	37th BS, 28th BW Ellsworth AFB
86-0103	96th BW Dyess AFB	7th BW Dyess AFB	9th BS, 7th BW Dyess AFB
86-0104	28th BW Ellsworth AFB	28th BW Ellsworth AFB	34th BS, 366th Wing Mountain Home AFB
86-0105	319th BW Grand Forks AFB	7th BW Dyess AFB	9th BS, 7th BW Dyess AFB
86-0106	319th BW Grand Forks AFB	7th BW Dyess AFB	Crashed November 30, 1992 while assigned to the 337th BS, 96th BombWing Dyess AFB
86-0107	319th BW Grand Forks AFB	116th BW GA ANG	128th BS, 116th BW GA ANG
86-0108	319th BW Grand Forks AFB	7th BW Dyess AFB	28th BS, 7th BW Dyess AFB
86-0109	319th BW Grand Forks AFB	7th BW Dyess AFB	28th BS, 7th BW Dyess AFB
86-0110	319th BW Grand Forks AFB	7th BW Dyess AFB	28th BS, 7th BW Dyess AFB
86-0111	319th BW Grand Forks AFB	28th BW Ellsworth AFB	77th BS, 28th BW Ellsworth AFB
86-0112	319th BW Grand Forks AFB	7th BW Dyess AFB	9th BS, 7th BW Dyess AFB
86-0113	319th BW Grand Forks AFB	28th BW Ellsworth AFB	37th BS, 28th BW Ellsworth AFB
86-0114	319th BW Grand Forks AFB	28th BW Ellsworth AFB	37th BS, 28th BW Ellsworth AFB
86-0115	319th BW Grand Forks AFB	184th BW KS ANG	127th BS, 184th BW KS ANG
86-0116	319th BW Grand Forks AFB	28th BW Ellsworth AFB	34th BS, 366th Wing Mountain Home AFB
86-0117	319th BW Grand Forks AFB	7th BW Dyess AFB	28th BS, 7th BW Dyess AFB
86-0118	319th BW Grand Forks AFB	28th BW Ellsworth AFB	34th BS, 366th Wing Mountain Home AFB
86-0119	319th BW Grand Forks AFB	7th BW Dyess AFB	28th BS, 7th BW Dyess AFB
86-0120	319th BW Grand Forks AFB	7th BW Dyess AFB	9th BS, 7th BW Dyess AFB
86-0121	319th BW Grand Forks AFB	28th BW Ellsworth AFB	34th BS, 366th Wing Mountain Home AFB
86-0122	319th BW Grand Forks AFB	7th BW Dyess AFB	28th BS, 7th BW Dyess AFB
86-0123	319th BW Grand Forks AFB	7th BW Dyess AFB	9th BS, 7th BW Dyess AFB
86-0124	384th BW McConnell AFB	116th BW GA ANG	128th BS, 116th BW GA ANG
86-0125	384th BW McConnell AFB	28th BW Ellsworth AFB	34th BS, 366th Wing Mountain Home AFB
86-0126	384th BW McConnell AFB	7th BW Dyess AFB	28th BS, 7th BW Dyess AFB
86-0127	384th BW McConnell AFB	184th BW KS ANG	127th BS, 184th BW KS ANG
86-0128	384th BW McConnell AFB	28th BW Ellsworth AFB	37th BS, 28th BW Ellsworth AFB
86-0129	384th BW McConnell AFB	28th BW Ellsworth AFB	37th BS, 28th BW Ellsworth AFB
86-0130	384th BW McConnell AFB	7th BW Dyess AFB	28th BS, 7th BW Dyess AFB
86-0131	319th BW Grand Forks AFB	28th BW Ellsworth AFB	77th BS, 28th BW Ellsworth AFB
86-0132	96th BW Dyess AFB	7th BW Dyess AFB	28th BS, 7th BW Dyess AFB
86-0133	28th BW Ellsworth AFB	28th BW Ellsworth AFB	128th BS, 116th BW GA ANG

86-0134	384th BW McConnell AFB	28th BW Ellsworth AFB	34th BS, 366th Wing Mountain Home AFB
86-0135	384th BW McConnell AFB	7th BW Dyess AFB	9th BS, 7th BW Dyess AFB
86-0136	384th BW McConnell AFB	184th BW KS ANG	127th BS, 184th BW KS ANG
86-0137	384th BW McConnell AFB	7th BW Dyess AFB	28th BS, 7th BW Dyess AFB
86-0138	384th BW McConnell AFB	28th BW Ellsworth AFB	34th BS, 366th Wing Mountain Home AFB
86-0139	384th BW McConnell AFB	28th BW Ellsworth AFB	34th BS, 366th Wing Mountain Home AFB
86-0140	384th BW McConnell AFB	7th BW Dyess AFB	9th BS, 7th BW Dyess AFB

B-1B AIRCRAFT

B-1B aircraft were manufactured in five production lots, not blocks as in the case of the B-52 and B-2. Each lot was funded by the U.S. Congress during a different budget year, and as a result the serial numbers of each production lot started with a different two digit number indicating the year in which production for those B-1s was funded. Lot I consisted of the B-1B production prototype (82-0001) only, Lot II included seven aircraft (83-0065 thru 83-0071), Lot III included ten aircraft (84-0049 thru 84-0058), Lot IV included 34 aircraft (85-0059 thru 85-0092) and Lot V included the last 48 aircraft (86-0093 thru 86-0140). The last four digits of Lot II aircraft are repeated in Lot IV, so there are two 0065s, 83-0065 and 85-0065; two 0066s, 83-0066 and 85-0066, etc. through 0071. The last three digits of Lots III, IV, and V serial numbers run consecutively 049 through 140, and as a result, the aircraft production number for these can be determined by subtracting 40 from the serial number (i.e. aircraft 84-0049 is the 9th production aircraft; aircraft 86-0140 is the 100th production aircraft, etc.) After transition to ACC, the old tactical style of tail number presentation was adopted displaying a small AF over the two digit year in small numbers, followed by the last three digits of the serial number in 30 inch high numbers. Unit Flagship aircraft violated this marking convention.

B-1B LOT I AIRCRAFT

82-0001 - LEADER OF THE FLEET

The prototype B-1B aircraft (82-0001) was hand built using major subassemblies which were planned to be used to build B-1A #5. Major items on #1 differed from similar items on the remainder of the B-1 fleet. The aircraft had full B-1B systems and a considerable amount of specialized flight test parameter monitoring systems were installed, including a pitot boom on the tip of the radome (which was later removed). It was delivered to the USAF on September 4, 1984. The aircraft was used at Edwards AFB as the primary test aircraft for aerodynamic and weapon release testing.

On August 22, 1988, the aircraft was flown to Ellsworth AFB, South Dakota and assigned to the 28th Bombardment Wing. The aircraft, being a prototype, required specialized equipment and procedures to operate. As a result of the high expense of operating this "one of a kind" prototype, it was removed from flight status and used at Ellsworth AFB as a weapons loading practice aircraft. The nickname "Leader Of The Fleet" was retained, with nose art of a knight on a horse added. During April and May 1995 the aircraft was disassembled and laid out on the alert pad, as seen in these photos, to allow the Russians to observe it dismantled. After a 90 day verification period, final disassembly was complete in August 1995. Most of the parts common to the other B-1Bs were removed to be used as spares to maintain the B-1 fleet.

These three photos show 82-0001 laid out on the alert pad in August 1995 for spy satellite conformation of its disassembly. (All three Author)

The major assemblies were used for live fire testing (actual anti-aircraft weapons were fired at the assemblies) to determine survivability of the aircraft. The damaged parts were then used to determine the best methods of battle damage repair. The "Leader of the Fleet" was the first B-1B to be removed from service, retiring with 617 flight hours accumulated in 138 test flights.

STAR OF ABILENE as the 7th Bomb Wing Flagship marked with the 9th and 28th Bomb Squadron Insignias. (Author)

STAR OF ABILENE as the 7th Wing Flagship marked with the 9th and 28th Bomb Squadron and 39th and 40th Airlift Squadron Insignias. (Alan Winzerling)

B-1B LOT II AIRCRAFT

83-0065 STAR OF ABILENE 7th BOMB WING FLAGSHIP

The aircraft was delivered to Edwards AFB and used for maintenance procedure (Technical Order) validation and verification in May and June of 1985. On June 17, 1985 the aircraft left Edwards and flew to Offutt AFB for delivery to SAC. While at Offutt AFB, the STAR OF ABILENE name and nose art was applied to both sides of the aircraft along with full color SAC and 96th Bomb Wing emblems. The STAR OF ABILENE was officially delivered to the 96th Bomb Wing at Dyess AFB on July 14, 1985.

On March 10, 1986, while on a training mission over Texas, the wing sweep mechanism failed with the wings set at a 65 degree sweep. The aircraft was flown from the Dyess AFB area to Edwards AFB to take advantage of the long Edwards runway. The aircraft landed with the wings in the 65 degree sweep position with a ground speed of 250 knots at touchdown. The aircraft stopped in under 9000 feet, with no damage (except hot brakes). The wing sweep problem was traced to a failure of a cable which senses the wing sweep position. After repair, the aircraft returned to Dyess. The failure was caused by improper installation of the cable. An inspection of the other B-1s was accomplished correcting any similar problems which were found.

On May 31, 1992, when 83-0065 was turned over to ACC, the aircraft had accumulated 1,240.8 flight hours. By August 31, 1996, 83-0065 was assigned to the 7th Bomb Wing and had accumulated 2,412.3 flight hours. It has kept the nickname STAR OF ABILENE, though the nose art has so far changed three times. As of August 1, 1998, it was assigned to the 9th Bomb Squadron, 7th Bomb Wing, and marked as the 7th Bomb Wing Flagship.

LEFT: STAR OF ABILENE as the 7th Bomb Wing Flagship marked with the 337th Bomb Squadron Insignia. (Author)

BELOW: 83-0065, marked as the 7th Wing Flagship is seen here on December 16, 1997, at Barksdale AFB. (Brian C. Rogers)

83-0066 OLE' PUSS

The aircraft was delivered to the 96th Bomb Wing at Dyess AFB, Texas on October 16, 1985. It was named OLE' PUSS and given the nose art carried in World War II on B-17F 42-30073 of the 96th Bomb Group stationed in England. A fuel tank explosion at Tinker AFB on May 26, 1989, blew a four foot hole in the side of the fuselage. The aircraft was repaired at Tinker AFB and returned to Dyess. As of May 31, 1992, when turned over to ACC, 83-0066 had accumulated 1,264.9 flight hours. It has kept the nickname OLE' PUSS, with the nose art remaining the same. By August 31, 1996, 83-0066 was assigned to the 7th Bomb Wing and had accumulated 2,034.4 flight hours. As of August 1, 1998, the aircraft was assigned to the 28th Bomb Squadron, 7th Bomb Wing at Dyess AFB.

(Author)

83-0066, OLE' PUSS, is seen here at Dyess AFB on August 13, 1993, wearing the blue and white checkered tail stripe of the 338th CCTS, 96th Bomb Wing. (Randy Walker)

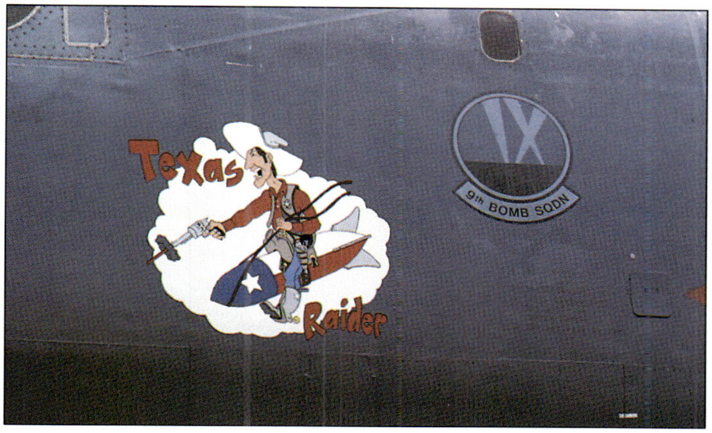

(Author)

83-0067 TEXAS RAIDER

The aircraft was delivered to the 96th Bomb Wing at Dyess AFB, Texas on January 13, 1986. It was given the name TEXAS RAIDERS and painted with nose art on both sides of the aircraft. The art was the same art carried on a Confederate Air Force B-17 also named TEXAS RAIDERS. The aircraft has been repainted in the all gunship gray scheme with no nose art. As of May 31, 1992, when turned over to ACC, 83-0067 had accumulated 1,137.4 flight hours. It has kept the nickname TEXAS RAIDER, with its nose art changed from the original. By August 31, 1996, 83-0067 was assigned to the 7th Bomb Wing and had accumulated 2,096.3 flight hours. As of August 1, 1998, the aircraft was assigned to the 9th Bomb Squadron, 7th Bomb Wing at Dyess AFB.

83-0067, TEXAS RAIDER, is seen here at Dyess AFB on February 10, 1995, wearing the black tail stripe of the 9th Bomb Squadron, 7th Bomb Wing. (Author)

83-0067 is seen here at Langley AFB on May 4, 1993, wearing the red tail stripe of the 337th Bomb Squadron, 96th Bomb Wing. (Brian C. Rogers)

83-0068 PREDATOR

The aircraft was delivered to the 96th Bomb Wing at Dyess AFB Texas on February 18, 1986. It was given the name SPUDS and painted with nose art of SPUDS McKENZIE, the Budweiser Beer mascot. As of May 31, 1992, when turned over to ACC, 83-0068 had accumulated 1,256.5 flight hours. While assigned to ACC, it carried the nickname PREDATOR, with the nose art of a black and white tiger. By August 31, 1996, 83-0068 was assigned to the 7th Bomb Wing and had accumulated 2,308.0 flight hours. As of August 1, 1998, the aircraft was assigned to the 28th Bomb Squadron, 7th Bomb Wing at Dyess AFB.

(Author)

83-0068, PREDATOR, is seen here at Dyess AFB wearing the blue and white checkered tail stripe of the 28th Bomb Squadron, 7th Bomb Wing. (Author)

83-0068 is seen here at Dyess AFB on May 12, 1997, taxiing for takeoff. (Author)

(Author)

83-0069 THE BEAST

The aircraft was delivered to the 96th Bomb Wing at Dyess AFB, Texas on March 13, 1986. It was given the name SILENT PENETRATER. 83-0069 was later named REBEL and painted with nose art from B-29 42-65338 of the 497th Bomb Group. As of May 31, 1992, when turned over to ACC, 83-0069 had accumulated 1,393.0 flight hours. It carries a new nickname THE BEAST with nose art of two paws with claws in shackles. By August 31, 1996, 83-0069 was assigned to the 7th Bomb Wing and had accumulated 2,322.0 flight hours. As of August 1, 1998, the aircraft was assigned to the 28th Bomb Squadron, 7th Bomb Wing at Dyess AFB.

83-0069, THE BEAST, is seen here at Dyess AFB on February 20, 1995, wearing the blue and white checkered tail stripe of the 28th Bomb Squadron, 7th Bomb Wing. (Author)

83-0070 7 WISHES

During April 1986 the aircraft was used at Edwards AFB for MK 82 500 pound bomb fit and ground weapon separation tests. The aircraft was delivered to the 96th Bomb Wing at Dyess AFB Texas on May 6, 1986. Because it was the seventh B-1B built, it was given the name 7 WISHES with associated nose art. As of May 31, 1992, when turned over to ACC, 83-0070 had accumulated 1,131.9 flight hours. It has retained the nickname 7 WISHES, but the nose art has been changed twice since being assigned to ACC. By August 31, 1996, 83-0070 was assigned to the 7th Bomb Wing and had accumulated 2,160.3 flight hours. As of August 1, 1998, the aircraft was assigned to the 28th Bomb Squadron, 7th Bomb Wing at Dyess AFB.

The present version of 7 WISHES is seen here, photographed on May 12, 1997. (Author)

This earlier version of 7 WISHES photographed on July 27, 1994, has the Genie from Walt Disney's Aladdin movie. (Author)

83-0070, 7 WISHES, is seen here at Dyess AFB on July 27, 1994, wearing the blue and white checkered tail stripe of the 28th Bomb Squadron, 7th Bomb Wing. (Author)

83-0071 SPITFIRE

The aircraft was delivered to the 96th Bomb Wing at Dyess AFB, Texas on May 24, 1986. It was given the name GRAND ILLUSION with nose art of a mermaid on a nuclear bomb. A fire on #3 engine caused by a first stage fan disintegration occurred in the pattern at Dyess on December 19, 1990. The aircraft landed at Dyess. The engine failure and fire caused major damage to the #3 engine bay. The aircraft received a temporary fix at Dyess and was flown to Palmdale in June 1991 for permanent repairs. The aircraft was returned to Dyess after the repair. As of May 31, 1992, when turned over to ACC, 83-0071 had accumulated 995.2 flight hours. It was renamed SPITFIRE with nose art of a black panther breathing fire. Two versions of the nose art have been used since 83-0071 was assigned to ACC. By August 31, 1996, 83-0071 was assigned to the 7th Bomb Wing and had accumulated 2,379.4 flight hours. As of August 1, 1998, the aircraft was assigned to the 9th Bomb Squadron, 7th Bomb Wing at Dyess AFB.

ABOVE LEFT: The present SPIT FIRE nose art seen here was photographed at Dyess AFB on May 1, 1998. (Author)

LEFT: This version of SPIT FIRE was photographed at Dyess AFB on February 20, 1995. (Author)

83-0071, SPIT FIRE, is seen here at Dyess AFB on February 20, 1995, wearing the black tail stripe of the 9th Bomb Squadron, 7th Bomb Wing. (Author)

This view of 83-0071 was taken at Dyess AFB on May 1, 1998, wearing the black tail stripe of the 9th Bomb Squadron, 7th Bomb Wing. (Author)

LOT III AIRCRAFT

84-0049 THUNDER FROM THE SKY

The aircraft was delivered to Edwards AFB to be used as the ALCM (AGM-86B) test aircraft, and was the first B-1 to incorporate the Long Intermediate Stores Bay modification. The aircraft had an "External Obvious Difference" (EOD) antenna mounted on the top of the aircraft. This antenna was a non functioning ALT-32H antenna (from the B-52), installed to comply with SALT II treaty restrictions The aircraft was converted back to the standard three bay configuration; however, the EOD antenna remained, indicating that 049 has the capability of being converted back to the cruise missile capable stores bay.

The aircraft carries the ED tail code of the 412th Test Wing. As of May 31, 1992 the aircraft had accumulated 694.8 flight hours. It carries the nickname, THUNDER FROM THE SKY, with the nose art of a gauntlet holding a lance. It is assigned to the 419th Test Squadron, 412th Test Wing at Edwards AFB. By August 31, 1996, 84-0049 had accumulated 1,108.3 flight hours.

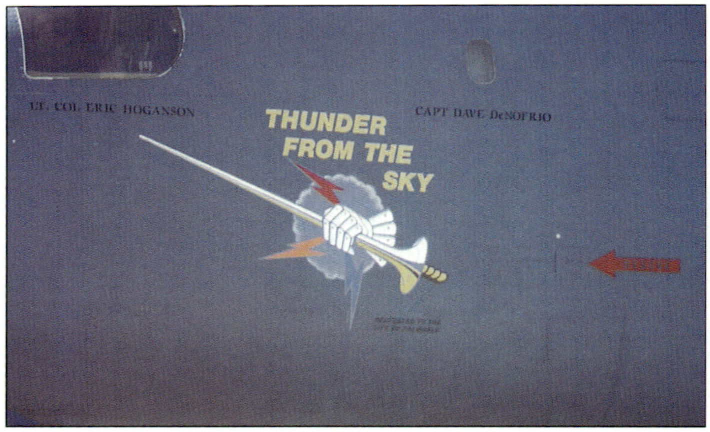

(Author)

BELOW: 84-0049, THUNDER FROM THE SKY, is seen here at Edwards AFB during April 1995 assigned to the 419th Test Squadron, 412th Test Wing. (Author)

84-0049 is seen here releasing a dummy JDAM as part of weapon separation testing at Edwards AFB. (USAF)

(Author)

84-0050

The aircraft was used for hot/cold weather testing at the Climatic Laboratory at Eglin AFB. The aircraft was delivered to the 96th Bomb Wing at Dyess AFB, Texas on October 1, 1986. It was initially given the name SURF RAT, with no nose art. During the summer of 1991, 84-0050 was renamed SURPRISE ATTACK with nose art of a girl with a towel, painted on both sides of the aircraft As of May 31, 1992, when turned over to ACC, 84-0050 had accumulated 1298.7 flight hours. By August 31, 1996, 84-0050 was assigned to the 7th Bomb Wing and had accumulated 1,990.0 flight hours. The aircraft presently does not have nose art. As of August 1, 1998, the aircraft was assigned to the 28th Bomb Squadron, 7th Bomb Wing at Dyess AFB.

84-0050 is seen here at Tinker AFB, Oklahoma on September 3, 1993, with the blue and white checkered tail stripe of the 338th CCTS, 96th Bomb Wing. (Author)

84-0051 BOSS HAWG

The aircraft was delivered to the 96th Bomb Wing at Dyess AFB, Texas on July 30, 1986. It was given the name LUCKY LADY with nose art of a girl in a bathing suit painted on both sides of the aircraft. During the summer of 1991, 84-0051 was renamed BOSS HAWG, and the nose art of an Arkansas razorback pig replaced the girl. As of May 31, 1992, when turned over to ACC, 84-0051 had the most flight hours of any B-1, having accumulated 2,166.4 flight hours. BOSS HAWG has retained its nickname and nose art. By August 31, 1996, 84-0051, assigned to the 7th Bomb Wing, had accumulated 3,389.1 flight hours. As of August 1, 1998, the aircraft was assigned to the 9th Bomb Squadron, 7th Bomb Wing at Dyess AFB.

(Author)

84-0051, BOSS HAWG, is seen here at Dyess AFB on February 20, 1995, with the black tail stripe of the 9th Bomb Squadron, 7th Bomb Wing. (Author)

84-0052 Aircraft Crashed on September 28, 1987, Before Change-Over To ACC

The aircraft was delivered to the 96th Bomb Wing at Dyess AFB, Texas on July 11, 1986. The aircraft crashed at La Junta, Colorado on September 28, 1987. Possible bird strike followed by an engine fire resulted in loss of the aircraft. The aircraft was carrying a crew of six. There were three successful ejections and three fatalities. The copilot's seat failed to eject leaving the Instructor pilot occupying the seat stranded in the aircraft along with two other crew members. There was not enough time to manually bailout through the bottom hatch before the aircraft hit the ground.

(Author)

84-0053 LUCKY 13

The aircraft was delivered to the 96th Bomb Wing at Dyess AFB, Texas on July 10, 1986. Because it was the thirteenth B-1B built, it was given the name LUCKY 13 with nose art of a leprechaun in front of a four leaf clover. The aircraft was sent to Palmdale in October of 1987 for ejection seat testing following the failure of the copilots ejection seat in the loss of aircraft 84-0052. As of May 31, 1992, when turned over to ACC, 84-0053 had accumulated 1,348.6 flight hours. It has retained the nickname LUCKY 13 and the nose art of a four leaf clover. The leprechaun has been removed. By August 31, 1996, 84-0053 was assigned to the 7th Bomb Wing and had accumulated 2,139.2 flight hours. As of August 1, 1998, the aircraft was assigned to the 28th Bomb Squadron, 7th Bomb Wing at Dyess AFB.

84-0053, LUCKY 13, the thirteenth B-1B built, is seen here at Dyess AFB on February 20, 1995, with the blue and white checkered tail stripe of the 28th Bomb Squadron, 7th Bomb Wing. (Author)

84-0054

The aircraft was delivered to the 96th Bomb Wing at Dyess AFB, Texas on July 26, 1986. It was given the name SILVER BULLET with no nose art. It was later named TASMANIAN TERROR with nose art of a Tasmanian Devil cartoon character throwing a bomb. As of May 31, 1992, when turned over to ACC, the aircraft had accumulated 1,584.4 flight hours. By August 31, 1996, 84-0054 was assigned to the 7th Bomb Wing and had accumulated 2,7925.6 flight hours. As of August 1, 1998, the aircraft was assigned to the 28th Bomb Squadron, 7th Bomb Wing at Dyess AFB and was not carrying nose art.

84-0054 is seen here taxiing at Dyess AFB on July 26, 1993, with the blue and white checkered tail stripe of the 338th CCTS, 96th Bomb Wing. (Author)

84-0055

The aircraft was delivered to the 96th Bomb Wing at Dyess AFB, Texas on August 22, 1986. It was given the name RIDGE RUNNER and nose art painted on both sides of the aircraft. During the summer of 1991, the 84-0055 was renamed SUNRISE SURPRISE and nose art of a Polynesian girl in front of a setting sun. As of May 31, 1992, when turned over to ACC, the aircraft had accumulated 1,926.9 flight hours. By August 31, 1996, 84-0055 was assigned to the 7th Bomb Wing and had accumulated 3,132.3 flight hours. As of August 1, 1998, the aircraft was assigned to the 28th Bomb Squadron, 7th Bomb Wing at Dyess AFB and was not carrying nose art.

84-0055 is seen here at Dyess AFB on February 20, 1995, with the blue and white checkered tail stripe of the 28th Bomb Squadron, 7th Bomb Wing. (Author)

(Author)

84-0056 SWEET SIXTEEN
28th BOMB SQUADRON FLAGSHIP

The aircraft was delivered to the 96th Bomb Wing at Dyess AFB, Texas on August 27, 1986. Because it was the sixteenth B-1B built, it was given the name SWEET SIXTEEN with nose art, on both sides of the aircraft, of a cartoon B-1 in front of a large red-orange circle. As of May 31, 1992, when turned over to ACC, 84-0056 had accumulated 1,782.6 flight hours. By August 31, 1996, 84-0056 was assigned to the 7th Bomb Wing and had accumulated 3,365.0 flight hours. It has retained the nickname and the nose art, and is assigned to the 28th Bomb Squadron, 7th Bomb Wing at Dyess AFB. It is marked as the 28th Bomb Squadron Flagship.

84-0056, SWEET 16, the sixteenth B-1B built, is seen here at Dyess AFB on February 20, 1995, marked as the 28th Bomb Squadron Flagship. (Author)

84-0056 is seen here at Tinker AFB on September 3, 1993, with the blue and white checkered tail stripe of the 338th CCTS, 96th Bomb Wing. (Author)

84-0057 THE HELLION
9th BOMB SQUADRON FLAGSHIP
Crashed February 18, 1998, near Mattoon, Kentucky

The aircraft was sent to Edwards during late 1986 and early 1987 to be used for Electro-Magnetic Interference/Radar Cross Section testing. The aircraft was then delivered to the 96th Bomb Wing at Dyess AFB, Texas on April 30, 1987. It was given the name THE HELLION with nose art, on both sides of the aircraft, of a heavy metal type eagle. As of May 31, 1992, when turned over to ACC, 84-0057 had accumulated 1,993.2 flight hours. By August 31, 1996, 84-0057 was assigned to the 7th Bomb Wing and had accumulated 3,473.4 flight hours. While assigned to ACC, it kept the nickname and updated nose art and was marked as the 9th Bomb Squadron Flagship.

84-0057 crashed February 18, 1997 at approximately 1:45 p.m. CST near Mattoon, Ky. The aircraft was on a training mission and was not carrying any munitions. The crew members were Lt. Col. Daniel J. Charchian, Pilot; Capt. Jeffrey Sabella, Co-pilot; Capt. Kevin J. Schields, Weapons Systems Officer; and 1st Lt. Bert G. Winslow, Weapons System Officer. All four aircrew members ejected safely and were treated for minor injuries. The cause was traced to a faulty fire control panel. When one engine was intentionally shut down using the control panel a short circuit in the panel shut down all four engines. With no hydraulic or electrical power, the aircraft became uncontrollable forcing the crew to eject.

(Author)

BELOW: 84-0057, HELLION, is seen here at Dyess AFB on February 20, 1995, marked as the 9th Bomb Squadron Flagship. These were the markings it was carrying when it crashed. (Author)

BOTTOM: 84-0057 is seen here at Tinker AFB on September 3, 1993, with the red tail stripe of the 337th Bomb Squadron, 96th Bomb Wing. (Author)

(Author)

84-0058 ETERNAL GUARDIAN

The aircraft was delivered to the 96th Bomb Wing at Dyess AFB, Texas on March 26, 1987. It was given the name MASTER OF DISASTER with nose art of a girl in front of a winged creature. As of May 31, 1992, when turned over to ACC, 84-0058 had accumulated 1,483.8 flight hours. It carries the nickname ETERNAL GUARDIAN with nose art. By August 31, 1996, 84-0058 was assigned to the 7th Bomb Wing and had accumulated 2,822.6 flight hours. As of August 1, 1998, the aircraft was assigned to the 9th Bomb Squadron, 7th Bomb Wing at Dyess AFB.

84-0058 is seen here at Tinker AFB on September 3, 1993, with the red tail stripe of the 337th Bomb Squadron, 96th Bomb Wing. (Author)

BELOW: 84-0058, ETERNAL GUARDIAN, is seen here at Nellis AFB during September 1995 with the black tail stripe of the 9th Bomb Squadron, 7th Bomb Wing. (Alan Winzerling)

84-0058 is seen here at Dyess AFB during December 30, 1993, with the black tail stripe of the 9th Bomb Squadron, 7th Bomb Wing. It also is marked with the white shadowed tail letters normally reserved for a unit Flagship. (Author)

LOT IV AIRCRAFT

85-0059 BAD DOG

The aircraft was delivered to the 96th Bomb Wing at Dyess AFB, Texas on September 26, 1986. It was given the name SUPER GLIDER with no nose art. During the summer of 1991, the aircraft was renamed BETTER DUCK. As of May 31, 1992, when turned over to ACC 85-0059 had accumulated 1,819.3 flight hours. It now carries the nickname BAD DOG and accompanying nose art. By August 31, 1996, 85-0059 was assigned to the 116th Bomb Wing and had accumulated 2,782.7 flight hours. As of August 1, 1998, the aircraft was assigned to the 128th Bomb Squadron, 116th Bomb Wing, Georgia Air National Guard at Robins AFB.

(Author)

85-0059 is seen here taxiing at Dyess AFB on July 27, 1994, with the blue and white checkered tail stripe of the 337th Bomb Squadron, 7th Bomb Wing. (Author)

85-0059, BAD DOG, is seen here at Dyess AFB on February 29, 1995, with the blue and white checkered tail stripe of the 28th Bomb Squadron, 7th Bomb Wing. (Alan Winzerling)

85-0060 REACH OUT & TOUCH SOMEONE

The aircraft was delivered to the 96th Bomb Wing at Dyess AFB, Texas on October 31, 1986. It was given the name NIGHT HAWK with nose art, on both sides of the aircraft, of a hawk in front of the moon. During early 1991, the aircraft was transferred to Ellsworth AFB and back to Dyess, It was re-named NIGHT HAWK. As of May 31, 1992, when turned over to ACC, 85-0060 had accumulated 1,612.7 flight hours. After assignment to ACC, while at Ellsworth, it was nicknamed ROLLING THUNDER with Buffalo nose art. By August 31, 1996, 85-0060 was assigned to the 184th Bomb Wing, Kansas Air National Guard and had accumulated 2,777.2 flight hours. Nose art of a combination Jayhawk and B-1 called REACH OUT AND TOUCH SOMEONE was added to the aircraft during late spring of 1998. As of August 1, 1998, the aircraft was assigned to the 184th Bomb Wing, Kansas Air National Guard.

ABOVE LEFT: (Author)
LEFT: (Randy Walker)

85-0060 is seen here at McConnell AFB on March 23, 1998, with the red and white tail stripe of the 127th Bomb Squadron, 184th Bomb Wing, Kansas Air National Guard (ANG). (Author)

85-0060, ROLLING THUNDER, is seen here at Ellsworth AFB on May 11, 1998, with the blue tail stripe of the 77th Bomb Squadron, 28th Bomb Wing. (Author)

85-0061

The aircraft was delivered to the 96th Bomb Wing at Dyess AFB, Texas on March 18, 1987. It was given the name FRENCH CONNECTION with no nose art. The aircraft was sent to the Paris Air Show in June 1987. During April 1991, the aircraft was transferred to Ellsworth AFB. As of May 31, 1992, when turned over to ACC, 85-0061 had accumulated 1,540.2 flight hours. After assignment to ACC, while at Ellsworth, it was nicknamed MAVERICK with wild horse nose art. It was repainted without the nose art being reapplied. By August 31, 1996, 85-0061 was assigned to the 28th Bomb Wing and had accumulated 2,506.4 flight hours. As of August 1, 1998, the aircraft was assigned to the 37th Bomb Squadron Tigers, 28th Bomb Wing at Ellsworth AFB.

(Author)

85-0061 is seen here at Ellsworth AFB on August 18, 1997, with the black and gold tail stripe of the 37th Bomb Squadron, 28th Bomb Wing. (Author)

85-0061, MAVERICK, is seen here at Ellsworth AFB on May 11, 1998, with the black and gold tail stripe of the 37th Bomb Squadron, 28th Bomb Wing. (Author)

85-0062 UNCAGED

The aircraft was delivered to the 96th Bomb Wing at Dyess AFB, Texas on November 4, 1986. It was given the name SKY DANCER with nose art of a girl riding on a winged horse. As of May 31, 1992, when turned over to ACC, 85-0062 had accumulated 1,747.2 flight hours. After reassignment, 85-0062 carried the nickname THE CHIEF with nose art of an Indian Chief for a short time until being renamed UNCAGED, with nose art of an eagle clawing its way through the fuselage skin. By August 31, 1996, 85-0062 was assigned to the 7th Bomb Wing and had accumulated 3,151.0 flight hours. As of August 1, 1998, the aircraft was assigned to the 9th Bomb Squadron, 7th Bomb Wing at Dyess AFB.

85-0063 Aircraft Crashed on November 8, 1988, Before Change-Over To ACC

The aircraft was delivered to the 96th Bomb Wing at Dyess AFB, Texas on November 30, 1986. It crashed at Dyess November 8, 1988 due to a fire in overwing fairing area. All four crew members ejected successfully.

ABOVE LEFT: (Author)
LEFT: (Alan Winzerling)

85-0062, UNCAGED, is seen here at Dyess AFB on February 20, 1995, with the black tail stripe of the 9th Bomb Squadron, 7th Bomb Wing. (Author)

85-0062, THE CHIEF, is seen here at Dyess AFB on September 30, 1993, with the red tail stripe of the 337th Bomb Squadron, 96th Bomb Wing. (Alan Winzerling)

85-0064
184TH BOMB WING FLAGSHIP

The aircraft was delivered to the 96th Bomb Wing at Dyess AFB, Texas on November 3, 1986. It was given the name ELIMINATOR with nose art on both sides of the aircraft, of a Ford hot rod with space shuttle type wings taken from a ZZ TOP record album. In December 1987, 85-0064 became the first B-1 to fly 100 sorties. The aircraft was reassigned to Ellsworth and the nose art was removed. As of May 31, 1992, when turned over to ACC, the aircraft had accumulated 1,445.7 flight hours. After assignment to ACC, while at Ellsworth, it was nicknamed PRAIRIE THUNDER with Buffalo head nose art. By August 31, 1996, 85-0064 was assigned to the 184th Bomb Wing and had accumulated 2,330.8 flight hours. It is assigned to the Kansas Air National Guard, and is marked as the 184th Bomb Wing Flagship.

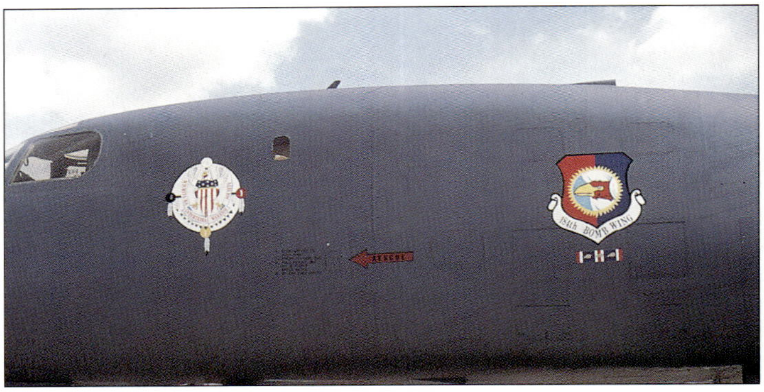

ABOVE RIGHT: (Alan Winzerling)
RIGHT: (Author)

85-0064 is seen here in April 1997 marked as the 184th Bomb Wing Flagship. (Douglas Slowiak)

85-0064, PRAIRIE THUNDER, is seen here at Ellsworth AFB on June 21, 1993, with the black and gold tail stripe of the 37th Bomb Squadron, 28th Bomb Wing. (Author)

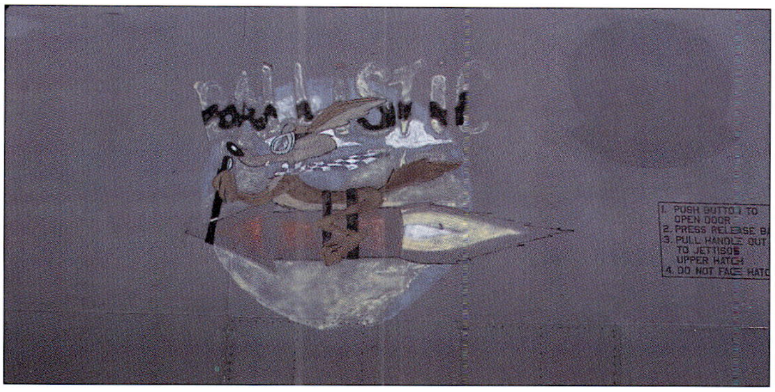

85-0065 BALLISTIC

The aircraft was delivered to the 96th Bomb Wing at Dyess AFB, Texas on November 30, 1986. It was given the name TRILOGY OF TERROR with nose art of a three headed dragon. It was renamed TEXAS ARMOR with nose art of an armadillo wearing tennis shoes holding a U.S. flag and a Texas flag painted on both sides of the aircraft. As of May 31, 1992, when turned over to ACC, 85-0065 has accumulated 1,315.6 flight hours. By August 31, 1996, 85-0065 was assigned to the 7th Bomb Wing and had accumulated 2,909.7 flight hours. It was renamed BALLISTIC with new nose art added. As of August 1, 1998, the aircraft was assigned to the 128th Bomb Squadron, 116th Bomb Wing, Georgia Air National Guard at Robins AFB.

ABOVE LEFT: (Nate Leong)
LEFT: (Author)

85-0065 is seen here at Robins AFB, Georgia on July 23, 1998. (Nate Leong)

85-0065, TEXAS ARMOR, is seen here at Dyess AFB on June 27, 1994, with the blue and white tail stripe of the 337th Bomb Squadron, 7th Bomb Wing. (Author)

85-0066 BADLANDS BOMBER

The aircraft was delivered to the 96th Bomb Wing at Dyess AFB, Texas on December 31, 1986. It was given the name SPECIAL DELIVERY with nose art of a girl with a cape with a World War II type national insignia. 85-0066 was renamed MIS BEHAVIN with nose art of a girl in a bathing suit painted on both sides of the aircraft. The aircraft was transferred to Ellsworth AFB in early 1991, renamed MISSOURI MISS, and the nose art was replaced. The MISSOURI MISS nose art was originally on a B-24 (44-41811) of the 380th Bomb Group, and later on an FB-111A (68-0247) of the 380th Bomb Wing. As of May 31, 1992, when turned over to ACC, 85-0066 had accumulated 1,593.7 flight hours. After assignment to ACC, while at Ellsworth, it was nicknamed DEADWOOD EXPRESS with nose art of a Stage Coach applied to the left side. It was repainted and renamed as the BADLANDS BOMBER with nose art of a B-1 applied to the left side. By August 31, 1996, 85-0066 was assigned to the 28th Bomb Wing and had accumulated 3,022.2 flight hours. As of August 1, 1998, the aircraft was assigned to the 37th Bomb Squadron Tigers, 28th Bomb Wing at Ellsworth AFB.

ABOVE RIGHT: (Author)
RIGHT: (Author)

85-0066, BADLANDS BOMBER, is seen here at Ellsworth AFB on August 18, 1997, with the black and gold tail stripe of the 37th Bomb Squadron, 28th Bomb Wing. (Author)

85-0066, DEADWOOD EXPRESS, is seen here at Ellsworth AFB on June 21, 1993, with the blue tail stripe of the 77th Bomb Squadron, 28th Bomb Wing. (Author)

(Author)

85-0067 ON DEFENSE

The aircraft was delivered to the 96th Bomb Wing at Dyess AFB, Texas on December 30, 1986. It was given the name WILD THANG with nose art on both sides of the aircraft. After the transfer of 85-0066 to Ellsworth, 85-0067 was renamed MIS BEHAVIN and was painted with the art of a girl in a bathing suit. On 85-0067 the nose art was changed slightly adding a blanket under the girl. As of May 31, 1992, when turned over to ACC, 85-0067 had accumulated 1,190.8 flight hours. After assignment to ACC, while at Dyess, it repainted and nicknamed ON DEFENSE with nose art of an eagle head in front of a circular section of the American flag. By August 31, 1996, 85-0067 was assigned to the 7th Bomb Wing and had accumulated 2,409.3 flight hours. As of August 1, 1998, the aircraft was assigned to the 28th Bomb Squadron, 7th Bomb Wing at Dyess AFB. 85-0067 took part in Operation Desert Fox, flying missions from Oman.

85-0067, ON DEFENSE, is seen here at Dyess AFB on May 12, 1997, with the blue and white checkered tail stripe of the 28th Bomb Squadron, 7th Bomb Wing. (Author)

85-0067 is seen here taxiing at Dyess AFB on July 23, 1993, with the blue and white checkered tail stripe of the 338th CCTS, 96th Bomb Wing. (Author)

85-0068 DRAGON'S FURY

The aircraft was delivered to Edwards AFB to be used as the ACM (AGM-129) test aircraft. It is the second B-1 to have the Long Intermediate Stores Bay modification accomplished. The aircraft had an "External Obvious Difference" EOD antenna mounted on the top of the aircraft. This antenna, same as the antenna installed on 84-0049, was a non functioning ALT-32H antenna from the B-52 and was installed to comply with SALT II treaty restrictions requiring cruise missile carrier aircraft to be visually identifiable. The aircraft was converted back to the standard three bay configuration; however, the EOD antenna was retained. As of May 31, 1992, 85-0068 had accumulated 763.2 flight hours. It now is named DRAGON'S FURY, with art of a Dragon flying in front of the Earth. By August 31, 1996, 85-0068 was assigned to the 412th Bomb Wing and had accumulated 1,736.3 flight hours. As of August 1, 1998, the aircraft was assigned to the 419th Test Squadron, 412 Test Wing.

(Author)

(Author)

85-0068, DRAGONS FURY is seen here releasing a dummy JDAM as part of weapon separation testing at Edwards AFB. (USAF)

(Author)

85-0069 HOME IMPROVEMENT

The aircraft was delivered to the 96th Bomb Wing at Dyess AFB, Texas on December 23, 1986. It was given the name DAISY MAE with nose art of a hillbilly girl. As of May 31, 1992, when turned over to ACC, the aircraft had accumulated 1205.0 flight hours. After assignment to ACC, at Dyess AFB, it was assigned to the Kansas Air National Guard, nicknamed HOME IMPROVEMENT, and had a caricature of a B-1 applied as nose art. By August 31, 1996, 85-0069 was assigned to the 184th Bomb Wing and had accumulated 2,475.8 flight hours. As of August 1, 1998, the aircraft was assigned to the 127th Bomb Squadron, 184th Bomb Wing, Kansas Air National Guard at McConnell AFB.

85-0069, HOME IMPROVEMENT, is seen here at McConnell AFB on March 23, 1998, with the subdued gray tail stripe of the 127th Bomb Squadron, 184th Bomb Wing, Kansas ANG. (Author)

85-0069 is seen here taxiing at Dyess AFB on June 26, 1993, with the blue and white checkered tail stripe of the 338th CCTS, 96th Bomb Wing. (Author)

85-0070

The aircraft was delivered to the 96th Bomb Wing at Dyess AFB, Texas on January 7, 1987. It was given the name EXCALIBUR with nose art of a girl's head and shoulders pointing a sword towards the nose. On October 4, 1989, after a hydraulic failure, the nose gear failed to extend due to a loose electrical plug in the emergency nose gear extension system. The aircraft made a landing, with only the main landing gear extended, on the shuttle landing strip at Edwards AFB, resulting in minor damage to the radome and radome bulkhead. As of May 31, 1992, when turned over to ACC, the aircraft had accumulated 1,386.0 flight hours. After assignment to ACC, at Dyess AFB, 85-0070 was reassigned to the Kansas Air National Guard, and repainted without nose art being applied. By August 31, 1996, 85-0062 was assigned to the 184th Bomb Wing and had accumulated 2,920.5 flight hours. As of August 1, 1998, the aircraft was assigned to the 127th Bomb Squadron, 184th Bomb Wing.

85-0070 is seen here at Dyess AFB on December 30, 1993, with the blue and white checkered tail stripe of the 337th Bomb Squadron, 96th Bomb Wing. (Author)

85-0070 is seen here at McConnell AFB on November 23, 1994, with the subdued gray tail stripe of the 127th Bomb Squadron, 184th Bomb Wing, Kansas ANG. (Author)

85-0071

The aircraft was delivered to the 28th Bomb Wing at Ellsworth AFB, South Dakota on February 4, 1987. The aircraft was later transferred to the 96th Bomb Wing at Dyess AFB, Texas. It was given the name LIBERATOR with associated nose art. As of May 31, 1992, when turned over to ACC, the aircraft had accumulated 2,041.9 flight hours. By August 31, 1996, 85-0071 was assigned to the 7th Bomb Wing and had accumulated 3,189.0 flight hours. As of August 1, 1998, the aircraft was assigned to the 128th Bomb Squadron 116th Bomb Wing, Georgia Air National Guard.

85-0071 is seen here at Dyess AFB on December 30, 1993, with the blue and white checkered tail stripe of the 337th Bomb Squadron, 96th Bomb Wing. (Author)

85-0071 is seen here at Robins AFB, Georgia on March 13, 1997, assigned to the 128th Bomb Squadron, 116th Bomb Wing, Georgia ANG. (Brian C. Rogers)

85-0072 POLARIZED

The aircraft was delivered to the 96th Bomb Wing at Dyess AFB, Texas on February 27, 1987. 85-0072 was given the name POLARIZED with nose art of a woman with a shield riding a polar bear. POLARIZED was originally used by AFOTEC at Dyess and had extra test equipment and radio beacons installed. On April 14, 1986, 85-0072 flew from Dyess, over the north pole, and along the northern border of the Soviet Union for a mission duration of 21.7 hours. POLARIZED was the first B-1 to participate in Red Flag June 22 through 26, 1987. POLARIZED flew the Distant Mariner Mission during August of 1988 from Dyess to Hickam AFB, Hawaii, then to Andersen AFB, Guam. While on Guam, the aircraft flew to Australia. As of May 31, 1992, when turned over to ACC, the aircraft had accumulated 1,955.9 flight hours. By August 31, 1996, 85-0072 was assigned to the 7th Bomb Wing and had accumulated 3,341.7 flight hours. As of August 1, 1998, the aircraft was assigned to the 9th Bomb Squadron, 7th Bomb Wing at Dyess AFB.

(Author)

85-0072, POLARIZED, is seen here at Dyess AFB on February 20, 1995, with the black stripe of the 9th Bomb Squadron, 7th Bomb Wing. (Author)

85-0072 is seen here Dyess AFB on December 30, 1993, with the black tail stripe of the 9th Bomb Squadron, 7th Bomb Wing. (Author)

85-0073 CERBERUS

The aircraft was delivered to the 28th Bomb Wing at Ellsworth AFB, South Dakota on January 13, 1987. It was the aircraft at the official B-1 arrival at Ellsworth, held on January 21, 1987. The aircraft was named WINGS OF FREEDOM and had nose art of a hawk's head in a white oval. 85-0073, was later transferred to the 96th Bomb Wing at Dyess AFB, Texas. It was given a new name with nose art from Greek mythology. 85-0073 now called CERBERUS, had nose art depicting the three headed dog that guarded the entrance to Hades. As of May 31, 1992, when turned over to ACC, the aircraft had accumulated 1,658.7 flight hours. The nose art subject remained the same but has been updated. By August 31, 1996, 85-0073 was assigned to the 7th Bomb Wing and had accumulated 3,402.5 flight hours. As of August 1, 1998, the aircraft was assigned to the 28th Bomb Squadron, 7th Bomb Wing at Dyess AFB.

(Author)

(Author)

85-0073, CERBERUS, is seen here at Dyess AFB on February 20, 1995, with the blue and white stripe of the 28th Bomb Squadron, 7th Bomb Wing. (Author)

85-0074 CREW DAWG

The aircraft was delivered to the 28th Bomb Wing at Ellsworth AFB, South Dakota on February 18, 1987. The aircraft was later transferred to the 96th Bomb Wing at Dyess AFB, Texas. It was given the name PENETRATOR and nose art of a warrior on a flying horse. 85-0074 had accumulated 1,614.8 flight hours when turned over to ACC on May 31, 1992. The aircraft was renamed CREW DAWG with nose art of four wolves, representing the four crew members, in front of the moon. By August 31, 1996, 85-0074 was assigned to the 7th Bomb Wing and had accumulated 2,993.2 flight hours. As of August 1, 1998, the aircraft was assigned to the 9th Bomb Squadron, 7th Bomb Wing at Dyess AFB.

(Author)

85-0074, CREW DAWG, is seen here at Dyess AFB on February 20, 1995, with the black tail stripe of the 9th Bomb Squadron, 7th Bomb Wing. (Brian C. Rogers)

85-0074, PENETRATOR, is seen here at Dyess AFB on December 26, 1993, with the black stripe of the 9th Bomb Squadron, 7th Bomb Wing. (Author)

85-0075 DAKOTA DEMOLITION

The aircraft was delivered to the 28th Bomb Wing at Ellsworth AFB, South Dakota on March 21, 1987. The aircraft was later transferred to the 96th Bomb Wing at Dyess AFB, Texas and then back to the 28th Bomb Wing at Ellsworth. It was given the name BANSHEE with nose art of a woman in blue in front of the moon. As of May 31, 1992, when turned over to ACC, the aircraft had accumulated 1,546.0 flight hours. By August 31, 1996, 85-0075 was assigned to the 28th Bomb Wing and had accumulated 3,059.8 flight hours. As of August 1, 1998, the aircraft was assigned to the 37th Bomb Squadron Tigers, 28th Bomb Wing at Ellsworth AFB.

(Author)

(Author)

85-0075, SPIRIT OF '76, is seen here at Ellsworth AFB on June 13, 1993, with the blue tail stripe of the 77th Bomb Squadron, 28th Bomb Wing. (Author)

85-0075, DAKOTA DEMOLITION, is seen here at Ellsworth AFB on August 18, 1997, with the black and gold tail stripe of the 37th Bomb Squadron, 28th Bomb Wing. (Author)

85-0076 BLACK JACK
Aircraft Crashed Before the Change-Over To ACC

The aircraft was delivered to the 28th Bomb Wing at Ellsworth AFB, South Dakota on March 19, 1987. It was given the name BLACK JACK with nose art of an Ace and Jack of Spades painted on the left side of the aircraft. The aircraft crashed, during marginal weather on final approach to Ellsworth on November 17, 1988.

85-0077 77th BOMB SQUADRON FLAGSHIP

The aircraft was delivered to the 28th Bomb Wing at Ellsworth AFB, South Dakota on February 13, 1987. 85-0077 was later transferred to the 96th Bomb Wing at Dyess AFB, Texas. It was given the name BONES (also called JAP HAPPY) with nose art from a World War II design of a skeleton in a leather flying helmet throwing a bomb. The nose art was painted on both sides of the aircraft. 85-0077 was transferred back to Ellsworth in 1990 and was named HAMTON. Nose art of a pig in an Air Force uniform in front of the American flag was added on the left side of the aircraft. As of May 31, 1992, when turned over to ACC, the aircraft had accumulated 1,376.9 flight hours. By August 31, 1996, 85-0077 was assigned to the 28th Bomb Wing and had accumulated 2,670.4 flight hours. As of August 1, 1998, the aircraft was assigned to the 77th Bomb Squadron, 28th Bomb Wing at Ellsworth AFB.

(Author)

BELOW: 85-0077, the 77th Bomb Squadron Flagship, is seen here at Ellsworth AFB on August 18, 1997, with the maroon and yellow tail stripe of the 77th Bomb Squadron, 28th Bomb Wing. (Author)

85-0077, PRIDE OF SOUTH DAKOTA, is seen here at Ellsworth AFB on June 13, 1994. It is marked as the 77th Bomb Squadron Flagship with the blue tail stripe of the 77th Bomb Squadron, 28th Bomb Wing. (Author)

(Author)

85-0078 Crashed September 19, 1997 near Alzada Montana

The aircraft was delivered to the 28th Bomb Wing at Ellsworth AFB, South Dakota on February 21, 1987. The aircraft was later transferred to the 96th Bomb Wing at Dyess AFB, Texas. It was given the name HEAVY METAL, with nose art of a sword through a shield, the shield having a lion painted on it. The aircraft was later reassigned to Ellsworth. As of May 31, 1992, when turned over to ACC, the aircraft had accumulated 1,407.3 flight hours. By August 31, 1996, 85-0078 was assigned to the 28th Bomb Wing and had accumulated 2,690.6 flight hours. The aircraft crashed on September 19, 1997, near Alzada, Montana. Col. Tony Beat, Major Clay Culver, Major Kirk Cakerice, and Captain Gary Everett were killed in the crash. The crash, the second B-1 lost after the change-over to ACC, was caused by pilot error. The aircraft did not have nose art at the time of the crash.

85-0077 is seen here at Ellsworth AFB on August 18, 1997, with the black and gold tail stripe of the 37th Bomb Squadron, 28th Bomb Wing. These were the markings carried by the aircraft when it crashed. (Author)

85-0078, DAKOTA LIGHTNING, is seen here at Ellsworth AFB on June 21, 1993, with the blue tail stripe of the 77th Bomb Squadron, 28th Bomb Wing. (Author)

85-0079 28th OPERATIONS GROUP FLAGSHIP

The aircraft was delivered to the 28th Bomb Wing at Ellsworth AFB, South Dakota on March 13, 1987. It was given the name WARRIORS DREAM, with nose art of a native girl under a waterfall. In the early spring of 1992, the aircraft was renamed CLASSY LADY with new nose art painted on the left side. As of May 31, 1992, when turned over to ACC, the aircraft had accumulated 1,704.5 flight hours. It carries art work of "The Deadman's Poker Hand" -Aces and Eights, and the name DEADWOOD DEALER. By August 31, 1996, 85-0079 was assigned to the 28th Bomb Wing and had accumulated 3,101.1 flight hours. It is marked as the 28th Operations Group Flagship and as of August 1, 1998, was assigned to the 37th Bomb Squadron Tigers, 28th Bomb Wing at Ellsworth AFB.

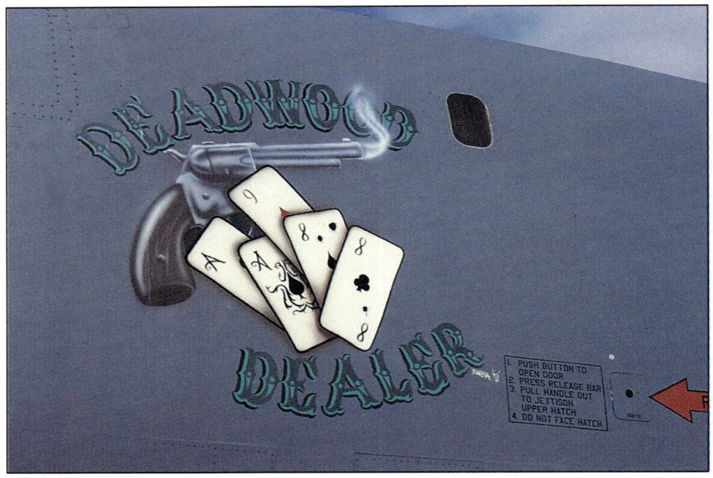

Present DEADWOOD DEALER nose art. (Author)

FAR LEFT: DEADWOOD DEALER nose art, photographed on June 13, 1994. (Author)

LEFT: CLASSY LADY nose art, photographed on July 27, 1992, was replaced by the more "politically correct" DEADWOOD DEALER. (Author)

85-0079, DEADWOOD DEALER, the 28th Operations Group Flagship, is seen here at Ellsworth AFB on August 18, 1997, with the black and gold tail stripe of the 37th Bomb Squadron, 28th Bomb Wing. (Author)

85-0079, CLASSY LADY, is seen here at Ellsworth AFB on July 27, 1992, with the blue tail stripe of the 77th Bomb Squadron, 28th Bomb Wing. (Author)

(Author)

85-0080

The aircraft was delivered to the 28th Bomb Wing at Ellsworth AFB, South Dakota on March 13, 1987. 0080 was given the name LADY OF THE NITE with nose art of a girl on a winged horse in front of the moon. 85-0080 cycled in and out of Edwards AFB between January 1988 through November 1989 and was used for ECM testing, including being placed on the turntable in the Air Force Anechoic Facility (AFAF). 85-0080 was delivered to the 384th Bomb Wing at McConnell AFB on July 20, 1990. It was given the name THE GATE KEEPER, with nose art of a Devil's face with bat like wings. The art was painted in red, outlined in black (the nose art is a copy of the squadron emblem of the 44th Fighter Squadron), and applied to the left side of the aircraft. The name GATE KEEPER had a copyright limiting its use. As a result, the name was changed to SCREAMIN DEMON and added to the nose art. As of May 31, 1992, when turned over to ACC, the aircraft had accumulated 1,005.7 flight hours. 85-0080 was reassigned to the Kansas Air National Guard, and repainted with nose art of WICHITA THUNDER (Wichita, Kansas minor league ice hockey team) being applied. By August 31, 1996, 85-0080 was assigned to the 184th Bomb Wing and had accumulated 2,119.2 flight hours. As of August 1, 1998, the aircraft was assigned to the 127th Bomb Squadron, 184th Bomb Wing at McConnell AFB.

BELOW: 85-0080, WICHITA THUNDER, is seen here at McConnell AFB on March 23, 1998, with the red and white tail stripe of the 127th Bomb Squadron, 184th Bomb Wing, Kansas ANG. (Author)

85-0080 is seen here at McConnell AFB on August 11, 1994, with the light blue KEEPER OF THE PLAINS tail stripe of the 28th Bomb Squadron, 384th Bomb Wing. (Author)

85-0081

The aircraft was delivered to the 28th Bomb Wing at Ellsworth AFB, South Dakota on April 1, 1987. It was given the name EQUALIZER, with nose art of a hooded executioner with an ax. The aircraft was transferred to Dyess AFB and then on February 1, 1991, to McConnell AFB. As of May 31, 1992, when turned over to ACC, the aircraft had accumulated 1,385.2 flight hours. 85-0081 was reassigned to the Kansas Air National Guard, and repainted without nose art being applied. By August 31, 1996, 85-0081 was assigned to the 184th Bomb Wing and had accumulated 2,243.4 flight hours. As of August 1, 1998, the aircraft was assigned to the 127th Bomb Squadron, 184th Bomb Wing, Kansas Air National Guard at McConnell AFB.

85-0081 is seen here at McConnell AFB on April 17, 1996, with the subdued gray tail stripe of the 127th Bomb Squadron, 184th Bomb Wing, Kansas ANG. (Author)

85-0081 is seen here at McConnell AFB on May 13, 1993, with the light blue KEEPER OF THE PLAINS tail stripe of the 28th Bomb Squadron, 384th Bomb Wing. (Author)

85-0082 GLOBAL POWER

The aircraft was delivered to the 28th Bomb Wing at Ellsworth AFB, South Dakota on March 23, 1987. It was given the name GUNSMOKE, with nose art of a six gun in front of an aces and eights (dead mans hand) poker hand. During spring of 1991, the aircraft was transferred to Dyess AFB and the same nose art was added to the right side of the aircraft. As of May 31, 1992, when turned over to ACC, the aircraft had accumulated 1,301.2 flight hours. Nose art of a dragon and the name TEXAS FIREPOWER were added. The nose art and name was changed to GLOBAL POWER with art of a bat in front of the globe. The nose art subject and name remained the same, but the art has been updated. By August 31, 1996, 85-0082 was assigned to the 7th Bomb Wing and had accumulated 2,836.3 flight hours. As of August 1, 1998, the aircraft was assigned to the 9th Bomb Squadron, 7th Bomb Wing at Dyess AFB. 85-0082 is currently assigned to the 412th Test Wing at Edwards AFB as the B-1 DSUP Test aircraft.

ABOVE LEFT: GLOBAL POWER nose art was photographed at Dyess AFB on May 12, 1997. (Author)

FAR LEFT: A more subdued GLOBAL POWER nose art was photographed at Dyess AFB on February 20, 1995. (Author)

LEFT: TEXAS FIREPOWER nose art was photographed at Dyess AFB on December 30, 1993. (Author)

85-0082, GLOBAL POWER, is seen here at Dyess AFB on February 20, 1995, with the black tail stripe of the 9th Bomb Squadron, 7th Bomb Wing. (Author)

85-0082 is seen here at Edwards AFB on October 2, 1998. (Keith Snyder)

85-0083 OVERNIGHT DELIVERY

The aircraft was delivered to the 28th Bomb Wing at Ellsworth AFB, South Dakota on March 30, 1987. It was given the name DARK STAR with nose art of a horse's head. As of May 31, 1992, when turned over to ACC, the aircraft had accumulated 1,462.1 flight hours. The aircraft was renamed OVERNIGHT DELIVERY with art of an eagle carrying a bomb. By August 31, 1996, 85-0083 was assigned to the 28th Bomb Wing and had accumulated 3,217.2 flight hours. As of August 1, 1998, the aircraft was assigned to the 37th Bomb Squadron Tigers, 28th Bomb Wing at Ellsworth AFB.

(Author)

85-0083, OVERNIGHT DELIVERY, is seen here at Ellsworth AFB on August 18, 1997, with the black and gold tail stripe of the 37th Bomb Squadron, 28th Bomb Wing. (Author)

85-0083, OVERNIGHT DELIVERY, is seen here at Ellsworth AFB on June 21, 1993, with the blue tail stripe of the 77th Bomb Squadron, 28th Bomb Wing. (Author)

(Brian C. Rogers)

85-0084 BRUTE FORCE

The aircraft was delivered to the 28th Bomb Wing at Ellsworth AFB, South Dakota on April 17, 1987. It was given the name PANDORAS BOX. Nose art of a woman in front of a treasure chest was painted on the left side of the aircraft. As of May 31, 1992, when turned over to ACC, the aircraft had accumulated 1,294.5 flight hours. The aircraft was renamed BRUTE FORCE with art of a tiger carrying a lightning bolt coming out of a globe. By August 31, 1996, 85-0084 was assigned to the 28th Bomb Wing and had accumulated 2,834.4 flight hours. As of August 1, 1998, the aircraft was assigned to the 37th Bomb Squadron Tigers, 28th Bomb Wing at Ellsworth AFB.

85-0084, BRUTE FORCE, is seen here at Nellis AFB on October 18, 1993, with the black and gold tail stripe of the 37th Bomb Squadron, 28th Bomb Wing. (Brian C. Rogers)

85-0085
The aircraft was delivered to the 28th Bomb Wing at Ellsworth AFB, South Dakota on May 4, 1987. It was given the name AMERICA #1 and nose art of a large letter A with an eagle painted on the left side of the aircraft. As of May 31, 1992, when turned over to ACC, the aircraft had accumulated 1,217.2 flight hours. The aircraft was repainted and its nose art was removed. By August 31, 1996, 85-0085 was assigned to the 7th Bomb Wing and had accumulated 2,959.6 flight hours. As of August 1, 1998, the aircraft was assigned to the 77th Bomb Squadron, 28th Bomb Wing at Ellsworth AFB.

(Author)

85-0085, AMERICA 1, is seen here at Ellsworth AFB on August 18, 1997, with the maroon and yellow tail stripe of the 77th Bomb Squadron, 28th Bomb Wing. (Author)

85-0085, AMERICA 1, is seen here at Ellsworth AFB on June 21, 1993, with the black and gold tail stripe of the 37th Bomb Squadron, 28th Bomb Wing. (Author)

85-0086 INTIMIDATOR

The aircraft was delivered to the 28th Bomb Wing at Ellsworth AFB, South Dakota on April 22, 1987. It was given the name MY MISTRESS with nose art of a girl with a bomb. The nose art was painted on both sides of the aircraft. This aircraft was the first B-1 on alert at Ellsworth on August 27, 1987. New nose art, also called MY MISTRESS, was added to the left side of the all gray B-1. As of May 31, 1992, when turned over to ACC, the aircraft had accumulated 1,758.2 flight hours. Nose art of a an eagle over a B-1 silhouette and the name SOARING WITH EAGLES were added. The nose art and name was changed to INTIMIDATOR with art of a mountain lion. By August 31, 1996, 85-0086 was assigned to the 28th Bomb Wing and had accumulated 3,316.4 flight hours. As of August 1, 1998, the aircraft was assigned to the 37th Bomb Squadron Tigers, 28th Bomb Wing at Ellsworth AFB.

ABOVE LEFT: INTIMIDATOR nose art was photographed at Ellsworth AFB on August 18, 1997. (Author)

LEFT: SOARING WITH EAGLES nose art was photographed at Ellsworth AFB on July 21, 1993. (Author)

BELOW: 85-0086, INTIMIDATOR, is seen here at Ellsworth AFB on August 18, 1997, with a somewhat sun bleached black tail stripe of the 37th Bomb Squadron, 28th Bomb Wing.

85-0086, SOARING WITH EAGLES, is seen here at Ellsworth AFB on June 21, 1993, with the black and gold tail stripe of the 37th Bomb Squadron, 28th Bomb Wing. (Author)

85-0087

The aircraft was delivered to the 28th Bomb Wing at Ellsworth AFB, South Dakota on May 1, 1987. It was given the name GREMLIN, with nose art of a gremlin with a hammer and wrench painted on the left side of the aircraft. As of May 31, 1992, when turned over to ACC, the aircraft had accumulated 1,347.5 flight hours. The aircraft was repainted and its nose art was removed. By August 31, 1996, 85-0087 was assigned to the 37th Bomb Squadron, 28th Bomb Wing and had accumulated 2,765.1 flight hours. New nose art of STARS AND STRIPES was added. In early 1997, the nose art was removed when 85-0087 was transferred to the 77th Bomb Squadron. As of August 1, 1998, the aircraft was assigned to the 77th Bomb Squadron, 28th Bomb Wing at Ellsworth AFB.

ABOVE RIGHT: STARS AND STRIPES nose art was carried on 85-0087, as seen here on June 13, 1994, until the aircraft transferred to the 37th Bomb Squadron in early 1997. (Author)

RIGHT: GREMLIN nose art was photographed on 85-0087 at Ellsworth AFB on July 27, 1992. (Author)

85-0087 is seen here deployed at Hill AFB on April 30, 1998, with the maroon and yellow tail stripe of the 77th Bomb Squadron, 28th Bomb Wing. (Brian C. Rogers)

85-0087, GREMLIN, is seen here at Ellsworth AFB on July 27, 1992, with the blue tail stripe of the 77th Bomb Squadron, 28th Bomb Wing. (Author)

85-0088 LOADED DICE

The aircraft was delivered to the 28th Bomb Wing at Ellsworth AFB, South Dakota on June 22, 1987. On February 9, 1988 the aircraft suffered a total electrical failure during a landing approach. 85-0088 was given the name PHOENIX and nose art of a Phoenix bird in front of a blue circle was painted on the left side of the aircraft. As of May 31, 1992, when turned over to ACC, the aircraft had accumulated 987.9 flight hours. After assignment to ACC, at Ellsworth AFB, it was assigned to the Kansas Air National Guard, and nicknamed LOADED DICE. By August 31, 1996, 85-0088 was assigned to the 184th Bomb Wing and had accumulated 2,239.6 flight hours. As of August 1, 1998, the aircraft was assigned to the 127th Bomb Squadron, 184th Bomb Wing, Kansas Air National Guard at McConnell AFB.

ABOVE LEFT: LOADED DICE was photographed on 85-0088 on November 23, 1994. (Author)

FAR LEFT: PHOENIX RISING was carried by 85-0088 while assigned to the 37th Bomb Squadron as seen here on July 17, 1993. (Randy Walker)

LEFT: PHOENIX was carried on the aircraft when it transferred to ACC. (Author)

85-0088, LOADED DICE, is seen here at Nellis AFB in November, 1995, with the subdued gray tail stripe of the 127th Bomb Squadron, 184th Bomb Wing, Kansas ANG. (David Brown)

85-0088, PHOENIX, is seen here at Ellsworth AFB on July 27, 1992, with the blue tail stripe of the 77th Bomb Squadron, 28th Bomb Wing. (Author)

85-0089 THE LAST PATROL

The aircraft was delivered to the 28th Bomb Wing at Ellsworth AFB, South Dakota on June 22, 1987. It was given the name MIDNIGHT PROWLER. Nose art of a tiger in front of the moon was painted on the left side of the aircraft. As of May 31, 1992, when turned over to ACC, the aircraft had accumulated 1,399.6 flight hours. The nose art name and subject remained the same, but the art was changed twice. By August 31, 1996, 85-0089 was assigned to the 28th Bomb Wing and had accumulated 2,961.4 flight hours. After transferring to the 128th Bomb Squadron, nose art of the POW/MIA emblem was added and the aircraft was given the name THE LAST PATROL. As of August 1, 1998, the aircraft was assigned to the 128th Bomb Squadron, 116th Bomb Wing, Georgia Air National Guard at Robins AFB.

(Nate Leong)

85-0089 is seen here at Robins AFB, Georgia on July 23, 1998. (Nate Leong)

ABOVE LEFT: The last version of MIDNIGHT PROWLER nose art, applied after the aircraft was repainted, was photographed on 85-0089 at Ellsworth AFB on July 27, 1992. CENTER: This version of MIDNIGHT PROWLER was applied to 85-0089 during 1987 and is seen here at Ellsworth AFB on July 27, 1992. RIGHT: After the transfer to ACC, the nose art was changed as seen here on June 21, 1993. (All Author)

85-0089, MIDNIGHT PROWLER, is seen here at Ellsworth AFB on June 21, 1993, with the blue tail stripe and EL tail code of the 28th Bomb Squadron. (Author)

(Author)

85-0090 TIGER COUNTRY

The aircraft was delivered to the 28th Bomb Wing at Ellsworth AFB, South Dakota on May 14, 1987. It was given the name TRAIL BLAZER. Nose art of a cowboy riding away was painted on the left side of the aircraft. The nose art on 85-0090 was later changed with the cowboy now riding towards you. As of May 31, 1992, when turned over to ACC, the aircraft had accumulated 1,067.7 flight hours. The nose art name was changed to TIGER COUNTRY and subject of the art was a tiger. By August 31, 1996, 85-0090 was assigned to the 28th Bomb Wing and had accumulated 2,391.4 flight hours. As of August 1, 1998, the aircraft was assigned to the 37th Bomb Squadron Tigers, 28th Bomb Wing at Ellsworth AFB.

85-0090, TIGER COUNTRY, is seen here at Ellsworth AFB on June 21, 1993, with the black and gold tail stripe of the 37th Bomb Squadron, 28th Bomb Wing. (Author)

85-0091

The aircraft was delivered to the 28th Bomb Wing at Ellsworth AFB, South Dakota on June 3, 1987. It was given the name THOR with nose art of the Norse god of thunder painted on the left side of the aircraft. The aircraft has been used for conventional weapons testing, including BDU-33 carriage/release testing (Short Trip 66) and carriage and release testing of inert MK-82 AIR 500 pound bombs. As of May 31, 1992, when turned over to ACC, the aircraft had accumulated 1,413.1 flight hours. The nose art name and subject were changed to MIDNIGHT STAR while assigned to the 28th Bomb Wing at Ellsworth AFB. By August 31, 1996, 85-0091 was assigned to the 28th Bomb Wing and had accumulated 2,603.5 flight hours. Nose art has been removed, and as of August 1, 1998, the aircraft was assigned to the 34th Bomb Squadron, 366th Wing at Mountain Home AFB.

MIDNIGHT STAR was carried on 85-0091 while it was assigned to the 28th Bomb Wing at Ellsworth AFB on June 13, 1994. (Author)

85-0091, MIDNIGHT STAR, is seen here at Offutt AFB during August 1993, with the blue tail stripe of the 28th Bomb Wing. (George Cockle)

85-0091 is seen here at Nellis AFB during September 1995 in the markings of the 34th Bomb Squadron, 366th Wing at Mountain Home AFB. (Author's Collection)

85-0092 THE UNINVITED

The aircraft was delivered to the 28th Bomb Wing at Ellsworth AFB, South Dakota on June 17, 1987. It was given the name ENFORCER, with nose art of a six gun with a sheriff's badge behind the gun pointing out of the aircraft. As of May 31, 1992, when turned over to ACC, the aircraft had accumulated 1,266.6 flight hours. After assignment to ACC, while at Ellsworth, it remained ENFORCER, then renamed THE UNINVITED with nose art of a tiger's head while assigned to the 37th Bomb Squadron. By August 31, 1996, 85-0092 was assigned to the 28th Bomb Wing and had accumulated 2,580.1 flight hours. It retained the nose art and as of August 1, 1998, was assigned to the 128th Bomb Squadron, 116th Bomb Wing, Georgia Air National Guard at Robins AFB.

ABOVE LEFT: (Nate Leong)
LEFT: (Author)

BELOW: 85-0092, THE UNINVITED, is seen here at Robins AFB, Georgia on March 13, 1997, assigned to the 128th Bomb Squadron, 116th Bomb Wing, Georgia ANG. (Brian C. Rogers)

85-0092 is seen here at Ellsworth AFB on June 21, 1993. While assigned to the 37th Bomb Squadron, 28th Bomb Wing, it carried nose art and the nickname ENFORCER. (Author)

LOT V AIRCRAFT

86-0093 GLOBAL POWER

The aircraft was delivered to the 28th Bomb Wing at Ellsworth AFB, South Dakota on July 27, 1987. It was given the name RUTHLESS RAVEN. The nose art is made up of a B-1 styled to look like a raven and a lightning bolt, painted on the left side of the aircraft. On March 24, 1992, 0093 was involved in a mid-air collision with KC-135R 62-3540 over Nebraska while practicing for SAC bomb competition. The collision occurred after the aircraft had completed air refueling. Both aircraft were badly damaged, but returned to Ellsworth with no injuries to the crew members. Both aircraft were repaired and returned to operational status. As of May 31, 1992, when turned over to ACC, the aircraft had accumulated 1,276.4 flight hours. The aircraft was renamed GLOBAL POWER with art of a tiger grasping the globe. By August 31, 1996, 86-0093 was assigned to the 28th Bomb Wing and had accumulated 2,866.7 flight hours. As of August 1, 1998, the aircraft was assigned to the 37th Bomb Squadron Tigers, 28th Bomb Wing at Ellsworth AFB.

(Author)

86-0093, GLOBAL POWER, is seen here at Ellsworth AFB on August 18, 1997, with black and gold rail stripe of the 37th Bomb Squadron, 28th Bomb Wing. (Author)

(Author)

86-0094

The aircraft was delivered to the 28th Bomb Wing at Ellsworth AFB, South Dakota on July 31, 1987. It was given the name NIGHT HAWK with nose art of a hawk's head. As of May 31, 1992, when turned over to ACC, the aircraft had accumulated 1,126.2 flight hours. The aircraft was repainted, and its nose art was removed. By August 31, 1996, 86-0094 was assigned to the 28th Bomb Wing and had accumulated 2,507.5 flight hours. As of August 1, 1998, the aircraft was assigned to the 77th Bomb Squadron, 28th Bomb Wing at Ellsworth AFB.

86-0094 is seen here at Ellsworth AFB on August 21, 1997, with the maroon and red tail stripe of the 77th Bomb Squadron, 28th Bomb Wing. (Author)

86-0094, NIGHT HAWK, is seen here at Ellsworth AFB on June 21, 1993, with the black and gold tail stripe of the 37th Bomb Squadron, 28th Bomb Wing. (Author)

86-0095

The aircraft was delivered to the 28th Bomb Wing at Ellsworth AFB, South Dakota on June 2, 1987. It was given the name MYSTIQUE with nose art of a high society woman. 86-0095 was transferred to Dyess AFB in early 1991 and renamed UNDECIDED with new nose art of a reclining woman replacing MYSTIQUE. As of May 31, 1992, when turned over to ACC, the aircraft had accumulated 1,591.5 flight hours. After assignment to ACC, at Dyess AFB, while assigned to the 337th Bomb Squadron, it was nicknamed TEXAS RATTLER with nose art of a rattlesnake. It was transferred to the 127th Bomb Squadron 184th Bomb Wing, Kansas Air National Guard, and the nose art was removed. By August 31, 1996, 86-0095 was assigned to the 184th Bomb Wing and had accumulated 2,709.2 flight hours. As of August 1, 1998, the aircraft was assigned to the Kansas Air National Guard at McConnell AFB.

(Brian C. Rogers)

86-0095 is seen here at McConnell AFB on March 23, 1998, with the subdued gray tail stripe of the 127th Bomb Squadron, 184th Bomb Wing, Kansas ANG. (Author)

86-0095, TEXAS RATTLER, is seen here at Langley AFB on February 3, 1994, with the blue and white checkered tail stripe of the 337th Bomb Squadron, 7th Bomb Wing. (Brian C. Rogers)

86-0096 WOLF PACK

The aircraft was delivered to the 28th Bomb Wing at Ellsworth AFB, South Dakota on July 2, 1987. It was given the name THUNDER CHILD and nose art of an Indian brave with a clenched fist and a lightning bolt. As of May 31, 1992, when turned over to ACC, the aircraft had accumulated 1,312.0 flight hours. The nose art and name was changed to WOLF PACK with art of a wolf's head. The nose art subject and name remained the same, but the art has been changed. By August 31, 1996, 86-0096 was assigned to the 28th Bomb Wing and had accumulated 2,378.8 flight hours. As of August 1, 1998, the aircraft was assigned to the 37th Bomb Squadron, 28th Bomb Wing at Ellsworth AFB. Taking part in Operation Desert Fox, 86-0096 was one of the first B-1s to see combat, flying on December 18, 1998 as SLAM 01.

The present version of WOLF PACK was photographed on August 21, 1997. (Author)

This version of WOLF PACK, photographed in May 1994, was removed when the aircraft was repainted. (Author)

86-0096, WOLF PACK, is seen here at Ellsworth AFB on August 21, 1997, with the black and gold tail stripe of the 37th Bomb Squadron, 28th Bomb Wing. (Author)

86-0097

The aircraft was delivered to the 28th Bomb Wing at Ellsworth AFB, South Dakota on June 25, 1987. It was given the name IRON EAGLE with nose art of an eagle with an American flag. During late 1991 the aircraft was transferred to Grand Forks AFB. As of May 31, 1992, when turned over to ACC, the aircraft had accumulated 1,237.4 flight hours. During 1994, the aircraft was transferred to Ellsworth AFB, and assigned to the 34th Bomb Squadron. It was repainted, and its nose art was removed. By August 31, 1996, 86-0097 was assigned to the 34th Bomb Squadron and had accumulated 2,578.9 flight hours. During 1997 it was reassigned to the 77th Bomb Squadron, 28th Bomb Wing at Ellsworth AFB.

The tail of 86-0097 is marked with the blue and orange tail stripe of the 319th Bomb Wing. The emblem in the stripe, the "Sunflake", is half sunburst and half snowflake. It was used on both SAC and ACC B-1s of the 319th Bomb Wing. (Author)

86-0097 is seen here, deployed at Hill AFB, on April 30, 1998, with the maroon and yellow tail stripe of the 77th Bomb Squadron, 28th Bomb Wing. (Brian C. Rogers)

86-0097 is seen here at Ellsworth AFB on June 21, 1993, with the blue and orange tail stripe of the 46th Bomb Squadron, 319th Bomb Wing, Grand Forks AFB, North Dakota.(Author)

86-0098 FREEDOM ONE

The aircraft was delivered to the 28th Bomb Wing at Ellsworth AFB, South Dakota on July 15, 1987. The aircraft broke 18 international speed and payload records on July 4, 1987. It was given the name FREEDOM I with nose art of a B-1 circling the globe. As of May 31, 1992, when turned over to ACC, the aircraft had accumulated 1,133.7 flight hours. After assignment to ACC, while assigned to the 37th Bomb Squadron at Ellsworth, it retained the name and art FREEDOM I. After the aircraft was repainted, the name remained the same, but the nose art was changed and "I" in the name was changed to "ONE." By August 31, 1996, 86-0098 was assigned to the 28th Bomb Wing and had accumulated 2,362.4 flight hours. It retained the nose art and as of August 1, 1998, was assigned to the 128th Bomb Squadron, 116th Bomb Wing, Georgia Air National Guard at Robins AFB.

ABOVE LEFT: The present version of FREEDOM ONE was photographed on March 13, 1997. (Brian C. Rogers)

LEFT: This version of FREEDOM I, photographed in June 21, 1993, was removed when the aircraft was repainted. (Author)

BELOW: 86-0098, FREEDOM ONE, is seen here at Robins AFB on March 13, 1997, in the markings of the 128th Bomb Squadron, 116th Bomb Wing, Georgia ANG. (Brian C. Rogers)

86-0098, FREEDOM I, is seen here at Ellsworth AFB on June 21, 1993, with the black and gold tail stripe of the 37th Bomb Squadron, 28th Bomb Wing. (Author)

86-0099 IRON EAGLE

The aircraft was delivered to the 28th Bomb Wing at Ellsworth AFB, South Dakota on August 3, 1987. The aircraft suffered a bird strike on September 19, 1990, heavily damaging both windscreens. It was given the name GHOST RIDER with nose art of a rider on horseback. 86-0099 was used for carriage and release testing of live MK82 AIR 500 pound bombs. GHOST RIDER completed nonnuclear certification of the B-1, on July 17, 1991, when it released 84 live MK82 AIR weapons on target at the Utah Test Range (UTTR) from an altitude of 500 feet AGL at a speed of .85 Mach with bomb spacing of .04 seconds (40 milliseconds) between bombs. As of May 31, 1992, when turned over to ACC, the aircraft had accumulated 1,491.2 flight hours. After assignment to ACC, while assigned to the 37th Bomb Squadron at Ellsworth, it was named IRON EAGLE with nose art of an eagle in front of the U.S. flag. After the aircraft was repainted, the name and art remained the same and was also repainted. By August 31, 1996, 86-0099 was assigned to the 28th Bomb Wing and had accumulated 2,873.7 flight hours. As of August 1, 1998, the aircraft was assigned to the 37th Bomb Squadron, 28th Bomb Wing at Ellsworth AFB.

ABOVE RIGHT: The present version of IRON EAGLE was photographed on August 18, 1997. (Author)

RIGHT: This version of IRON EAGLE, photographed in June 21, 1993, was removed when the aircraft was repainted. (Author)

BELOW: 86-0099, IRON EAGLE, is seen here at Ellsworth AFB on August 18, 1997, in the markings of the 37th Bomb Squadron, 28th Bomb Wing. (Author)

86-0099, IRON EAGLE, is seen here at Ellsworth AFB on June 21, 1993, with the blue tail stripe of the 28th Bomb Wing. (Author)

The present version of PHOENIX was photographed on May 12, 1997, while 86-0100 was assigned to the 28th Bomb Squadron, 7th Bomb Wing. (Author)

86-0100 PHOENIX

The aircraft was delivered to the 28th Bomb Wing at Ellsworth AFB, South Dakota on July 20, 1987. It was given the name PHANTOM with nose art of a stylized B-1 with a dragon's tail. 86-0100 was transferred to Dyess AFB in early 1991. The aircraft was renamed NIGHT HAWK, and new nose art was added. The new nose art is similar to that which had been on 85-0060, also NIGHT HAWK. As of May 31, 1992, when turned over to ACC, the aircraft had accumulated 1,723.7 flight hours. In July 1992, 86-0100 was reassigned to Dyess AFB. During 1994, the aircraft was grounded as a result of an electrical fire. It was repaired and the name PHOENIX with nose art of a Phoenix bird added. By August 31, 1996, 86-0100 was assigned to the 7th Bomb Wing and had accumulated 2,580.6 flight hours. As of August 1, 1998, the aircraft was assigned to the 28th Bomb Squadron, 7th Bomb Wing at Dyess AFB.

86-0100 is seen here at Dyess AFB on July 27, 1994, with the blue and white checkered tail stripe of the 337th Bomb Squadron, 7th Bomb Wing. (Author)

86-0101

The aircraft was delivered to the 28th Bomb Wing at Ellsworth AFB, South Dakota on August 21, 1987. It was given the name IRON BUTTERFLY with nose art of a stylized butterfly. The name was changed to LOW LEVEL DEVIL with nose art of a B-1 with a stylized devil's head, flying low level by making holes through the mountains. On December 2, 1991, 86-0101 was transferred to the 384th Wing at McConnell. As of May 31, 1992, when turned over to ACC, the aircraft had accumulated 1,262.7 flight hours. With the inactivation of the 384th Bomb Group, 86-0101 was transferred to Dyess AFB. It was repainted and nose art was not added. By August 31, 1996, 86-0101 was assigned to the 7th Bomb Wing and had accumulated 2,385.8 flight hours. As of August 1, 1998, the aircraft was assigned to the 28th Bomb Squadron, 7th Bomb Wing at Dyess AFB.

86-0101 is seen here at Dyess AFB on May 12, 1997, with the blue and white checkered tail stripe of the 28th Bomb Squadron, 7th Bomb Wing. (Author)

85-0101 is seen here at McConnell AFB on September 15, 1993, with the light blue KEEPER OF THE PLAINS tail stripe of the 28th Bomb Squadron, 384th Bomb Wing. (Author)

86-0102 BLACK HILLS SENTINEL

The aircraft was delivered to the 28th Bomb Wing at Ellsworth AFB, South Dakota. By August 31, 1987. It was given the name LADY HAWK with nose art, a woman's head with a hawk's head blended into her hair. As of May 31, 1992, when turned over to ACC, the aircraft had accumulated 1,344.8 flight hours. The nose art and name was changed to BLACK HILLS SENTINEL with art of a wolf howling at the moon. The aircraft was repainted with the name and nose art subject remaining the same, but the art was changed. By August 31, 1996, 86-0102 was assigned to the 28th Bomb Wing and had accumulated 2,576.8 flight hours. As of August 1, 1998, the aircraft was assigned to the 37th Bomb Squadron, 28th Bomb Wing at Ellsworth AFB. Flyinf out of Oman, 86-0102 saw combat during Operation Desert Fox, flying on December 19, 1998 as SLAM 04.

LEFT ABOVE: This version of BLACK HILLS SENTINEL was photographed on June 14, 1994. (Author)

LEFT: This earlier version of BLACK HILLS SENTINEL was photographed on September 3, 1993. (Author)

BELOW: 86-0102 is seen here at Ellsworth AFB on July 24, 1995, with the black and gold tail stripe of the 37th Bomb Squadron, 28th Bomb Wing. (Author)

85-0102, BLACK HILLS SENTINEL, is seen here at Ellsworth AFB on June 13, 1993, with blue tail stripe of the 28th Bomb Wing. (Author)

86-0103 RELUCTANT DRAGON

The aircraft was delivered to the 28th Bomb Wing at Ellsworth AFB, South Dakota on September 8, 1987. It was given the name HUNTRESS. Nose art of a woman hunter with a saber tooth tiger on each side was painted on the left side of the aircraft. 86-0103 was transferred to Dyess AFB in early 1991. The aircraft was renamed LOVELY LADY, and nose art of a girl in a pink bathing suit was added to the aircraft. As of May 31, 1992, when turned over to ACC, the aircraft had accumulated 1,802.2 flight hours. While assigned to the 337th Bomb Squadron, the nose art and name was changed to RELUCTANT DRAGON with art of a dragon holding a bomb. The aircraft was repainted and reassigned to the 9th Bomb Squadron, with the name and nose art subject remaining the same; but the art was changed. By August 31, 1996, 86-0103 was assigned to the 7th Bomb Wing and had accumulated 3,500.5 flight hours. As of August 1, 1998, the aircraft was assigned to the 9th Bomb Squadron, 7th Bomb Wing at Dyess AFB.

ABOVE RIGHT: This version of RELUCTANT DRAGON nose art was photographed on December 30, 1993.

RIGHT: This earlier version of RELUCTANT DRAGON nose art was photographed on April 25, 1993. (Alan Winzerling)

BELOW: 86-0103, RELUCTANT DRAGON, is seen here at Dyess AFB on December 30, 1993, with the black tail stripe of the 9th Bomb Squadron, 7th Bomb Wing. (Author)

85-0103 is seen here taxiing at Dyess AFB on August 13, 1993, with red tail stripe of the 337th Bomb Squadron, 96th Bomb Wing. (Randy Walker)

(Author)

86-0104

The aircraft was delivered to the 28th Bomb Wing at Ellsworth AFB, South Dakota on August 24, 1987. It was given the name AMERICAN FLYER. Nose art of an eagle with an American flag and arrows in its talons was painted on the left side of the aircraft. As of May 31, 1992, when turned over to ACC, the aircraft had accumulated 1,363.2 flight hours. During 1994, the aircraft was repainted and its nose art was removed. During 1995, it was reassigned to the 34th Bomb Squadron, 366th Wing while remaining at Ellsworth AFB. By August 31, 1996, 86-0104 was assigned to the 34th Bomb Squadron and had accumulated 2,600.5 flight hours. As of August 1, 1998, the aircraft was assigned to the 34th Bomb Squadron Tigers, 366th Wing at Mountain Home AFB.

86-0104 is seen here at Ellsworth AFB on July 24, 1995, with the black and red tail stripe of the 34th Bomb Squadron, 366th Wing. (Author)

85-0104, AMERICAN FLYER, is seen here during August 1992 with blue tail stripe of the 28th Bomb Wing. (Mike Grove)

86-0105 SNAKE EYES

86-0105 was the last B-1 delivered to Ellsworth, arriving on September 24, 1987. It was given the name SNAKE EYES with nose art of a cobra with a pair of dice. The aircraft was transferred to the 319th Bomb Wing at Grand Forks AFB during 1990. As of May 31, 1992, when turned over to ACC, the aircraft had accumulated 1,277.7 flight hours. During 1995, 86-0105 was transferred to Dyess AFB. It was repainted and the nose art was also repainted. By August 31, 1996, 86-0105 was assigned to the 7th Bomb Wing and had accumulated 2,730.8 flight hours. As of August 1, 1998, the aircraft was assigned to the 28th Bomb Squadron, 7th Bomb Wing at Dyess AFB.

(Author)

86-0105, SNAKE EYES, is seen here at Dyess AFB on May 12, 1997, with the blue and white checkered tail stripe of 28th Bomb Squadron, 7th Bomb Wing. (Author)

86-0106 LONE WOLF Crashed November 30, 1992

The aircraft was delivered to the 28th Bomb Wing at Ellsworth AFB, South Dakota on September 5, 1987. The aircraft was sent to Kirtland AFB in October 1987 for Electro-Magnetic Pulse (EMP) testing on the trestle at Kirtland AFB. It was given the name LONE WOLF with nose art of a wolf's head. 86-0106 was transferred to Dyess in late 1991. It retained the name LONE WOLF and its nose art. As of May 31, 1992, when turned over to ACC, the aircraft had accumulated 1,687.5 flight hours. On November 30, 1992, while assigned to the 337th Bomb Squadron, 96th Bomb Wing, 86-0106 crashed due to crew error. This occurred while on a night, low-level training mission near Van Horn, Texas. LONE WOLF was the first bomber lost after transition to ACC.

86-0106, LONE WOLF, is seen here at Dyess AFB on October 4, 1992, two months before it crashed. It is marked with the red tail stripe of the 337th Bomb Squadron, 96th Bomb Wing. (Author)

(Author)

86-0107 BAD TO THE BONE

The aircraft was delivered to the 319th Bomb Wing at Grand Forks AFB, North Dakota on October 26, 1937, six weeks early. It was given the name VINDICATOR. The nose art was painted on the left side of the aircraft and made up of a Norse warrior on a horse carrying a shield with 319 OMS (319th Organizational Maintenance Squadron) written on the shield. The name on 86-0107 was changed to VALKYRIES with new nose art of a woman with a torch riding on a horse. As of May 31, 1992, when turned over to ACC, the aircraft had accumulated 780.0 flight hours. During 1994, after assignment to Dyess AFB, it was renamed BAD TO THE BONE, and new nose art was added. The art was a skull, in a flight helmet, and cross bones in front of a B-1. By August 31, 1996, 86-0103 was assigned to the 116th Bomb Wing and had accumulated 1,895.3 flight hours. It retained the nose art and as of August 1, 1998, was assigned to the 128th Bomb Squadron, 116th Bomb Wing, Georgia Air National Guard at Robins AFB.

86-0107, BAD TO THE BONE, is seen here at Robins AFB on March 13, 1997, assigned to the 128th Bomb Squadron, 116th Bomb Wing, Georgia ANG. (Brian C. Rogers)

86-0107, BAD TO THE BONE, is seen here at Dyess AFB on February 20, 1995, with the black tail stripe of the 9th Bomb Squadron, 7th Bomb Wing. (Author)

86-0108 ALIEN WITH AN ATTITUDE

The aircraft was delivered to the 319th Bomb Wing at Grand Forks AFB, North Dakota on November 30, 1987. It was given the name HAWK with nose art of a cartoon hawk holding a nuclear bomb. As of May 3ˉ, 1992, when turned over to ACC, the aircraft had accumulated 1,053.8 flight hours. During 1994, after assignment to Dyess AFB, it was renamed ALIEN WITH AN ATTITUDE, and new nose art was added. By August 31, 1996, 86-0108 was assigned to the 7th Bomb Wing and had accumulated 2,490.0 flight hours. As of August 1, 1998, the aircraft was assigned to the 28th Bomb Squadron, 7th Bomb Wing at Dyess AFB.

(Author)

86-0108, ALIEN WITH AN ATTITUDE, is seen here at Langley AFB on August 11, 1994, assigned to the 337th Bomb Squadron, 7th Bomb Wing, Dyess AFB. (Brian C. Rogers)

86-0108, ALIEN WITH AN ATTITUDE, is seen here at Dyess AFB on May 1, 1998, with the blue and white checkered tail stripe of the 28th Bomb Squadron, 7th Bomb Wing. (Autho-)

(Author)

86-0109 SPECTRE - 7th OPERATIONS GROUP FLAGSHIP

The aircraft was delivered to the 319th Bomb Wing at Grand Forks AFB, North Dakota on October 1, 1987. 86-0109 was transferred to the 28th Bomb Wing at Ellsworth AFB, South Dakota on January 20, 1988, and given the name SPECTRE with nose art of a hooded skeleton. 86-0109 was transferred to the 96th Bomb Wing at Dyess AFB, Texas during the fall of 1990, and the SPECTRE nose art was added to the right side. As of May 31, 1992, when turned over to ACC, the aircraft had accumulated 1,436.4 flight hours. After the aircraft was repainted, the name remained the same, but the nose art was changed slightly. By August 31, 1995, 86-0109 was assigned to the 7th Bomb Wing and had accumulated 3,051.4 flight hours. As of August 1, 1998, the aircraft was marked as the 7th Operations Group (OG) Flagship and assigned to the 28th Bomb Squadron, 9th Bomb Wing at Dyess AFB. 86-0109 took part in Operation Desert Fox flying missions from Oman.

86-0109, SPECTRE, marked as the 7th Operations Group Flagship, is seen here taxiing at Dyess AFB on May 1, 1998, while assigned to the 28th Bomb Squadron, 7th Bomb Wing. (Author)

86-0109, SPECTRE, is seen here on September 25, 1992, with the blue and white checkered tail stripe of the 338th CCTS, 96th Bomb Wing. (Keith Snyder)

86-0110 STAIRWAY TO HEAVEN

The aircraft was delivered to the 319th Bomb Wing at Grand Forks AFB, North Dakota on September 19, 1987. The aircraft broke 18 international speed and payload records on September 17, 1987 (different records from those claimed by 86-0098 on July 4, 1987). It was given the name SUNRISE SURPRISE. Nose art of a cowgirl with a six gun in front of the sun was painted on the left side of the aircraft. The aircraft was Grand Forks AFB's first all gray B-1. As of May 31, 1992, when turned over to ACC the aircraft had accumulated 1,382.8 flight hours. During 1994 it was assigned to Dyess AFB. It was renamed STAIRWAY TO HEAVEN, and new nose art was added. By August 31, 1996, 86-0110 was assigned to the 7th Bomb Wing and had accumulated 2,799.8 flight hours. As of August 1, 1998, the aircraft was assigned to the 28th Bomb Squadron, 7th Bomb Wing at Dyess AFB.

(Author)

86-0110, STAIRWAY TO HEAVEN, is seen here at Dyess AFB on May 1, 1998, with the blue and white checkered tail stripe of the 28th Bomb Squadron, 7th Bomb Wing. (Author)

86-0110 is seen here at Grand Forks AFB on June 30, 1992, with the blue and orange tail stripe of the 46th Bomb Squadron, 319th Bomb Wing. (Author)

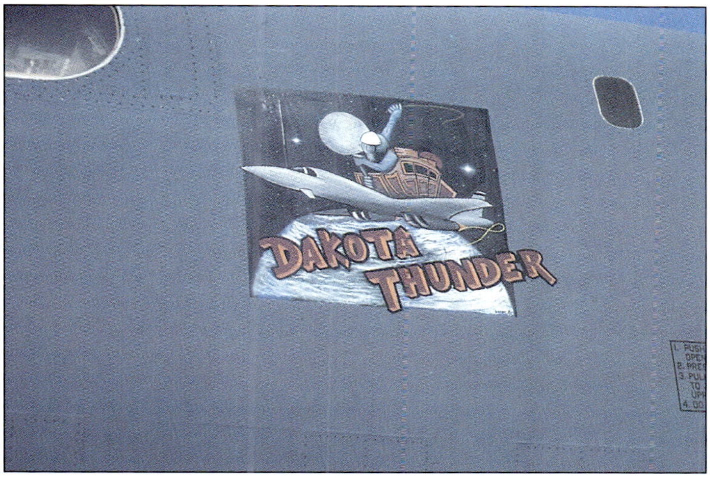

DAKOTA THUNDER is 86-0111's latest nose art photographed at Ellsworth AFB on August 17, 1997. (Author)

86-0111 DAKOTA THUNDER

The aircraft was delivered to the 319th Bomb Wing at Grand Forks AFB, North Dakota on October 16, 1987. It was given the name ACE IN THE HOLE with nose art of three aces up with one card down. During February and March 1992, 86-0111 and 86-0121 accomplished the flights which set twelve time-to-climb records for aircraft in the B-1 weight class. As of May 31, 1992, when turned over to ACC, the aircraft had accumulated 1,608.0 flight hours. During spring of 1994, the aircraft was transferred to Ellsworth, named PRAIRIE FIRE, and new nose art was added to the right side of the aircraft. In 1995, it was assigned to the 9th Bomb Squadron at Dyess for a short time. Upon returning to Ellsworth in 1996, the nose art and name was changed to DAKOTA THUNDER with art of a B-1 carrying a stagecoach. By August 31, 1996, 86-0111 was assigned to the 28th Bomb Wing and had accumulated 3,179.3 flight hours. As of August 1, 1998, the aircraft was assigned to the 37th Bomb Squadron, 28th Bomb Wing at Ellsworth AFB.

FAR LEFT: PRAIRIE FIRE nose art was on 86-0111 when photographed at Ellsworth AFB on June 13, 1994. (Author)

LEFT: ACE IN THE HOLE was carried on 86-0111 while assigned to the 319th Bomb Wing. (Author)

BELOW: 86-0111, DAKOTA THUNDER, is seen here at Ellsworth AFB on August 17, 1997, with the black and gold tail stripe of the 37th Bomb Squadron, 28th Bomb Wing. (Author)

86-0111, ACE IN THE HOLE, is seen here at Grand Forks AFB on June 30, 1992, with the blue and orange tail stripe of the 46th Bomb Squadron, 319th Bomb Wing. (Author)

86-0112 BLACK WIDOW

The aircraft was delivered to the 319th Bomb Wing at Grand Forks AFB, North Dakota on October 28, 1987. It was given the name VANNA. The nose art was made up of a woman in an "Indiana Jones" costume with hat, pistol and whip standing between two vine covered columns with the moon in the background. This art was painted on the left side of the aircraft. As of May 31, 1992, when turned over to ACC, the aircraft had accumulated 1,241.5 flight hours. The aircraft remained at Grand Forks AFB and was repainted with the nose art removed. During 1994, with the inactivation of the 319th Bomb Group, 86-0112 was assigned to the 9th Bomb Squadron at Dyess AFB. The nose art and name was changed to BLACK WIDOW. By August 31, 1996, 86-0112 was assigned to the 7th Bomb Wing and had accumulated 2,712.5 flight hours. As of August 1, 1998, the aircraft was assigned to the 9th Bomb Squadron, 7th Bomb Wing at Dyess AFB.

(Author)

86-0112, BLACK WIDOW, is seen here at Dyess AFB on February 20, 1995, with the black tail stripe of the 9th Bomb Squadron, 7th Bomb Wing. (Author)

86-0112 is seen here at Nellis AFB on October 18, 1993, with the blue and orange tail stripe of the 46th Bomb Squadron, 319th Bomb Wing, Grand Forks AFB. (Brian C. Rogers)

86-0113 DELIVERANCE

The aircraft was delivered to the 319th Bomb Wing at Grand Forks AFB, North Dakota on November 17, 1987. It was given the name CHARON with no nose art. The name was changed to VIPER with nose art of a snake wrapped around the name VIPER. 0113 was transferred to Ellsworth, and as of May 31, 1992, when turned over to ACC, the aircraft had accumulated 1,043.5 flight hours. During spring of 1994, the aircraft was transferred to Ellsworth, named DAKOTA REVEILLE, and new nose art was added to the right side of the aircraft. In 1996, the nose art and name was changed to DELIVERANCE. By August 31, 1996, 86-0113 was assigned to the 28th Bomb Wing and had accumulated 2,216.8 flight hours. During 1997, 86-0113 was modified with the AN/ALE-50 towed decoy system, flying test missions from Robins AFB, Georgia. As of August 1, 1998, the aircraft was assigned to the 37th Bomb Squadron, 28th Bomb Wing at Ellsworth AFB.

ABOVE LEFT: DELIVERANCE, is seen on 86-0113 as of May 19, 1996, but was removed by August 1997. (Author)

LEFT: DAKOTA REVEILLE was on 86-0113 when photographed at Ellsworth AFB on June 13, 1994. (Author)

BELOW: 86-0113 is seen here at Ellsworth AFB on August 21, 1997, with the black and gold tail stripe of the 37th Bomb Squadron, 28th Bomb Wing. (Author)

86-0113 is seen here at Ellsworth AFB on June 13, 1994, with the black and gold tail stripe of the 37th Bomb Squadron, 28th Bomb Wing. (Author)

86-0114 SCREAMIN EAGLE 37th BOMB SQUADRON FLAGSHIP

The aircraft was delivered to the 319th Bomb Wing at Grand Forks AFB, North Dakota on December 9, 1987. It was given the name WOLFHOUND with nose art of a large wolfhound dog with a studded collar. In May of 1989 and again in May of 1990, WOLFHOUND deployed to Mildenhall Air Base in England for the annual air show. On June 26, 1990, the aircraft suffered minor damage due to an inflight tail ECM compartment fire. The aircraft was repaired and brought back on flying status. As of May 31, 1992, when turned over to ACC, the aircraft had accumulated 1,240.8 flight hours. During spring of 1994, the aircraft was transferred to Ellsworth named DAKOTA DRIFTER, and new nose art was added to the right side of the aircraft. In 1996 the nose art and name was changed to SCREAMIN EAGLE. By August 31, 1996, 86-0114 was assigned to the 28th Bomb Wing and had accumulated 2,401.1 flight hours. As of August 1, 1998, the aircraft was assigned to the 28th Bomb Wing and marked as the 37th Bomb Squadron Flagship.

SCREAMIN EAGLE, 86-0114's latest nose art, was photographed on August 18, 1997. (Author)

DAKOTA DRIFTER was on 86-0114 when photographed at Ellsworth AFB on June 13, 1994. (Author)

86-0114, SCREAMIN EAGLE marked as the 37th Bomb Squadron Flagship is seen here at Ellsworth AFB on August 21, 1997. (Author)

86-0115 is seen here at McConnell AFB on August 21, 1992, with an eroded KEEPER OF THE PLAINS tail stripe. (Author)

86-0115

The aircraft was delivered to the 319th Bomb Wing at Grand Forks AFB, North Dakota on November 5, 1987. It was given the name BUMP AND RUN with no nose art. The name was changed to TOP SECRET with nose art of a sultry cartoon woman in red. During early 1992 the aircraft was transferred to McConnell AFB. As of May 31, 1992, when turned over to ACC, the aircraft had accumulated 810.9 flight hours. After assignment to ACC, at McConnell AFB, 86-0115 was reassigned to the Kansas Air National Guard, and repainted without nose art being applied. By August 31, 1996, 86-0115 was assigned to the 184th Bomb Wing and had accumulated 1,971.9 flight hours. As of August 1, 1998, the aircraft was assigned to the 127th Bomb Squadron, 184th Bomb Wing.

86-0115 is seen here at McConnell AFB on April 17, 1996, with the subdued gray tail stripe of the 127th Bomb Squadron, 184th Bomb Wing, Kansas ANG. (Author)

86-0115 is seen here at McConnell AFB on November 23, 1994, still painted in the SAC three color strategic camouflage. This color scheme used flat black tail markings including the stripe of the 127th Bomb Squadron, 184th Bomb Wing, Kansas ANG. (Author)

86-0116

The aircraft was delivered to the 319th Bomb Wing at Grand Forks AFB, North Dakota on December 2, 1987. It was given the name VICTRESS carrying nose art of woman in cavegirl skins. The art, on the left side of 86-0116, was later changed to a woman in a white bathing suit with a red cape and blue sash. As of May 31, 1992, when turned over to ACC, the aircraft had accumulated 837.3 flight hours. During 1994, with the inactivation of the 319th Bomb Group, 86-0116 was assigned to the 34th Bomb Squadron at Ellsworth. The aircraft was repainted and the nose art was removed. By August 31, 1996, 86-0116 was assigned to the 34th Bomb Squadron and had accumulated 2,310.9 flight hours. As of August 1, 1998, the aircraft was assigned to the 34th Bomb Squadron, 366th Wing at Mountain Home AFB.

BELOW: 86-0116 is seen beginning its takeoff roll as part of an exercise with the 34th Bomb Squadron deployed to Elmendorf AFB, Alaska, as part of Operation Amalgam Warrior during April 1996. (Dave Pasquini)

BELOW CENTER: 86-0116 is seen here at Ellsworth AFB on August 21, 1997, with the MO tail code of the 34th Bomb Squadron, 366th Wing. (Author)

86-0116 is seen here at Nellis AFB on October 19, 1993, with the blue and orange tail stripe of the 46th Bomb Squadron, 319th Bomb Wing, Grand Forks AFB. (Brian C. Rogers)

NIGHT STALKER was photographed on May 12, 1997. (Author)

86-0117 NIGHT STALKER

The aircraft was delivered to the 319th Bomb Wing at Grand Forks AFB North Dakota on December 4, 1987. It was given the name MILLENNIUM FALCON and had no nose art. The name was changed to PRIDE OF NORTH DAKOTA, and nose art of the Dakota Centennial seal was added on the left side of the aircraft. As of May 31, 1992, when turned over to ACC, the aircraft had accumulated 1,197.4 flight hours. During 1994, it was assigned to Dyess AFB. It was renamed NIGHT STALKER, and new nose art was added. By August 31, 1996, 86-0117 was assigned to the 7th Bomb Wing and had accumulated 2,871.9 flight hours. As of August 1, 1998, the aircraft was assigned to the 28th Bomb Squadron, 7th Bomb Wing at Dyess AFB.

86-0117, NIGHT STALKER, is seen here at Dyess AFB on May 12, 1997, with the blue and white checkered tail stripe of the 28th Bomb Squadron, 7th Bomb Wing. (Author)

86-0117 is seen here at Dyess AFB on July 27, 1994, in the three color strategic camouflage with the blue and white checkered tail stripe of the 28th Bomb Squadron, 7th Bomb Wing. (Author)

86-0118

The aircraft was delivered to the 319th Bomb Wing at Grand Forks AFB, North Dakota on November 18, 1987. The aircraft was accepted by the Air Force with no write-ups (deficiencies) on its first flight. It was given the name IRON MISTRESS with nose art of a woman warrior riding on a bear painted on the left side of the aircraft. IRON MISTRESS was repainted with an improved version of the woman and the polar bear. As of May 31, 1992, when turned over to ACC, the aircraft had accumulated 1,248.5 flight hours. During 1994, with the inactivation of the 319th Bomb Group, 86-0112 was assigned to the 34th Bomb Squadron at Ellsworth. The aircraft was repainted and the nose art was removed. By August 31, 1996, 86-0118 was assigned to the 34th Bomb Squadron and had accumulated 2,565.6 flight hours. As of August 1, 1998, the aircraft was assigned to the 34th Bomb Squadron, 366th Wing at Mountain Home AFB.

This nose art, similar to POLARIZED on 85-0072 was carried on 86-0118 while it was assigned to the 319th Bomb Wing at Grand Forks AFB. (Author)

86-0118 is seen here taxiing at Grand Forks AFB on July 30, 1992. (Author)

86-0118 is seen here at Ellsworth AFB on June 13, 1994, with the MO tail code of the 34th Bomb Squadron, 366th Wing, Mountain Home AFB. (Author)

86-0119

The aircraft was delivered to the 319th Bomb Wing at Grand Forks AFB, North Dakota on December 2, 1987. It was given the name SPUD with no nose art. The name was changed to CHRISTINE with nose art of a girl sitting on the hood of a red '57 Chevy with a license plate with the word EMUX (Electronic Multiplex System). During late 1991, the aircraft was transferred to Ellsworth AFB. As of May 31, 1992, when turned over to ACC, the aircraft had accumulated 859.1 flight hours. During 1994, it was assigned to Dyess AFB. It was repainted, and the nose art was removed. By August 31, 1996, 86-0119 was assigned to the 7th Bomb Wing and had accumulated 1,963.8 flight hours. It was flown at Edwards AFB during 1998 to test aircraft flight characteristics and flying qualities with intermediate wing sweep angles. It is assigned to the 28th Bomb Squadron, 7th Bomb Wing at Dyess AFB.

BELOW: 86-0119 is seen here at Dyess AFB on April 19, 1997, with the blue and white checkered tail stripe of the 28th Bomb Squadron, 7th Bomb Wing.(Author)

BELOW CENTER: 86-0119 is seen here at Dyess AFB on July 27, 1994, in the three color strategic camouflage with the blue and white checkered tail stripe of the 28th Bomb Squadron, 7th Bomb Wing. (Author)

86-0119 is seen here at Ellsworth AFB on July 30, 1992, with the blue and orange tail stripe of the 46th Bomb Squadron, 319th Bomb Wing, Grand Forks AFB, North Dakota. (Author)

86-0120 IRON HORSE

The aircraft was delivered to the 319th Bomb Wing at Grand Forks AFB, North Dakota on December 22, 1987. It was given the name MAD DAWG with nose art of a large cartoon dog holding a machine gun. As of May 31, 1992, when turned over to ACC, the aircraft had accumulated 1,369.3 flight hours. During 1994, with the inactivation of the 319th Bomb Group, 86-0120 was assigned to the 9th Bomb Squadron at Dyess. The aircraft was repainted and renamed IRON HORSE with new nose art. By August 31, 1996, 86-0120 was assigned to the 7th Bomb Wing and had accumulated 2,712.7 flight hours. As of August 1, 1998, the aircraft was assigned to the 9th Bomb Squadron, 7th Wing at Dyess AFB.

(Author)

86-0120, IRON HORSE, is seen here at Dyess AFB on May 12, 1997, with the black tail stripe of the 9th Bomb Squadron, 7th Bomb Wing. (Author)

BELOW: 86-0120 is seen here at Dyess AFB on July 27, 1994, in the three color strategic camouflage with the black tail stripe of the 9th Bomb Squadron, 7th Bomb Wing. A rotary launcher loaded with inert SRAM nuclear missiles is in the foreground in front of the nose landing gear. (Author)

MAD DAWG nose art was on 86-0120 during its assignment to the 319th Bomb Wing. (Author)

86-0120 is seen here at Grand Forks AFB on July 30, 1992, with the GF tail code of the 46th Bomb Squadron, 319th Bomb Wing. (Author)

86-0121

The aircraft was delivered to the 319th Bomb Wing at Grand Forks AFB, North Dakota on January 5, 1988. It was given the name EXTERMINATOR, later changed to TERMINATOR with no nose art. The aircraft was sent to Edwards AFB during January 1989 for bottom bailout system testing. The crew entry hatch and ladder were removed, and the nose gear retraction system was deactivated to simulate the condition of the aircraft when configured for bottom bailout. After 86-0121 returned to Grand Forks, the name was changed to ZEPPELIN with nose art of a hooded figure carrying a lantern (from a LED ZEPPELIN record album cover) painted on the left side of the aircraft. During February 1992, the name was changed to MAIDEN AMERICA. Nose art of a girl in a brown leather flight jacket with the words "Time To Climb" on the back of the jacket replaced the figure on the left side of 0121. "Time To Climb" referred to the flights which set twelve time to climb records for aircraft in the B-1 weight class during February and March 1992. 86-0111 and 86-0121 were the two aircraft that accomplished these record breaking flights. As of May 31, 1992, when turned over to ACC, the aircraft had accumulated 1,093.2 flight hours. During 1994, with the inactivation of the 319th Bomb Group, 86-0121 was assigned to the 34th Bomb Squadron at Ellsworth. The aircraft was repainted, and the nose art was removed. By August 31, 1996, 86-0121 was assigned to the 34th Bomb Squadron and had accumulated 2,394.1 flight hours. As of August 1, 1998, the aircraft was assigned to the 34th Bomb Squadron, 366th Wing at Mountain Home AFB.

(Author)

86-0121 is seen here at Ellsworth AFB on July 13, 1995, with the black and red tail stripe and MO tail code of the 34th Bomb Squadron, 366th Wing, Mountain Home AFB. (Author)

86-0121 is seen here at Grand Forks AFB on July 30, 1992, with the blue and orange tail stripe of the 46th Bomb Squadron, 319th Bomb Wing, Grand Forks AFB, North Dakota. (Author)

86-0122
The aircraft was delivered to the 319th Bomb Wing at Grand Forks AFB, North Dakota on January 12, 1988. It was given the name EXCALIBUR with nose art of a dragon and sword painted on the left side of the aircraft. As of May 31, 1992, when turned over to ACC, the aircraft had accumulated 1,184.9 flight hours. During 1994, with the inactivation of the 319th Bomb Group, 86-0112 was assigned to the 337th Bomb Squadron at Dyess. The aircraft was repainted and renamed RIDGE RUNNER with new nose art. By August 31, 1996, 86-0122 was assigned to the 7th Bomb Wing and had accumulated 2,442.0 flight hours. As of August 1, 1998, the aircraft was assigned to the 28th Bomb Squadron, 7th Wing at Dyess AFB.

86-0122, RIDGE RUNNER, is seen here at Dyess AFB on August 17, 1997, with the blue and white checkered tail stripe and of the 28th Bomb Squadron, 7th Bomb Wing. (Alan Winzerling)

86-0121 is seen here at Dyess AFB on July 27, 1994, in the three color strategic camouflage with the blue and white checkered tail stripe and of the 28th Bomb Squadron, 7th Bomb Wing. (Author)

86-0123
The aircraft was delivered to the 319th Bomb Wing at Grand Forks AFB, North Dakota on January 8, 1988. It was given the name MOLESTER with no nose art; the "MO" was later dropped leaving the name LESTER. The name was changed to HIGH NOON with nose art of a western bandit's head added to the aircraft. As of May 31, 1992, when turned over to ACC, the aircraft had accumulated 1,453.7 flight hours. During 1994, it was assigned to Dyess AFB. It was repainted, and the nose art was removed. By August 31, 1996, 86-0123 was assigned to the 7th Bomb Wing and had accumulated 2,874.0 flight hours. It is assigned to the 9th Bomb Squadron, 7th Bomb Wing at Dyess AFB.

86-0123 is seen here at Dyess AFB on February 20, 1995, in the three color strategic camouflage with the black tail stripe of the 9th Bomb Squadron, 7th Bomb Wing. (Author)

86-0123 is seen here at Langley AFB on February 5, 1993, with the blue and orange tail stripe of the 46th Bomb Squadron, 319th Bomb Wing, Grand Forks AFB, North Dakota. (Brian C. Rogers)

86-0124 GEORGIA GUARDIAN
116th BOMB WING FLAGSHIP

The aircraft was the first B-1 delivered to the 384th Bomb Wing at McConnell AFB, Kansas. It was delivered on January 4, 1988. The arrival nose art consisted of an eagle head and lightning bolt on top of a blue triangle painted on the left side of the aircraft. The nose art was removed in summer of 1988, and 86-0124 was given the name PENETRATOR, painted in an arc on the left side of the aircraft. During early 1989, the original nose art was repainted on the aircraft, but with its colors changed to more subdued shades. The aircraft was renamed WINGED THUNDER. As of May 31, 1992, when turned over to ACC, the aircraft had accumulated 1,210.7 flight hours. After assignment to ACC, it remained at McConnell until the 384th Bomb Group was inactivated. It moved to Dyess during 1995. While at Dyess, 86-0124 was repainted, and the nose art was removed. By August 31, 1996, 86-0124 was assigned to the 116th Bomb Wing and had accumulated 2,539.3 flight hours. It was named GEORGIA GUARDIAN and marked as the 116th Bomb Wing Flagship. As of August 1, 1998, the aircraft was assigned to the 128th Bomb Squadron, 116th Bomb Wing, Georgia Air National Guard at Robins AFB.

(Nate Leong)

86-0124 is seen here on June 13, 1997, assigned to the 128th Bomb Squadron, 116th Bomb Wing, Georgia ANG. (Jim Geer)

86-0124 is seen here at McConnell AFB on May 13, 1993. It was the first B-1 assigned to the 384th Bomb Wing at McConnell AFB. (Author)

86-0125

The aircraft was delivered to the 384th Bomb Wing at McConnell AFB, Kansas on February 4, 1988. It was given the name SHACK ATTACK which was painted in red in an arc on both sides of the aircraft with no nose art. Nose art was approved in June of 1992, but was never painted on the aircraft. As of May 31, 1992, when turned over to ACC, the aircraft had accumulated 1,267.1 flight hours. During 1994, with the inactivation of the 384th Bomb Group, 86-0125 was assigned to the 34th Bomb Squadron at Ellsworth. By August 31, 1996, 86-0125 was assigned to the 34th Bomb Squadron and had accumulated 2 381.3 flight hours. As of August 1, 1998, the aircraft was assigned to the 34th Bomb Squadron, 366th Wing at Mountain Home AFB.

BELOW: 86-0125 is seen here at McConnell AFB on December 2, 1992, with the light blue KEEPER OF THE PLAINS tail stripe and OZ tail code of the 28th Bomb Squadron, 384th Bomb Wing. (Author)

BELOW CENTER: 86-0125 is seen here at Ellsworth AFB on July 24, 1995, with the black and red tail stripe and MO tail code of the 34th Bomb Squadron, 366th Wing, Mountain Home AFB. (Author)

86-0125 is seen here at Dyess AFB on March 28, 1996, with the black and red tail stripe and MO tail code of the 34th Bomb Squadron, 366th Wing, Mountain Home AFB. (Alan Winzerling)

86-0126 HUNGRY DEVIL

The aircraft was delivered to the 384th Bomb Wing at McConnell AFB, Kansas on February 3, 1988. The arrival nose art painted on 86-0124 was also painted on this aircraft as a backup for the official arrival ceremony. The name on 86-0126 was later changed to COMMAND DECISION. Several nose art designs to go with the name COMMAND DECISION were suggested, but none were approved. As of May 31, 1992, when turned over to ACC, the aircraft had accumulated 1,188.8 flight hours. During 1994, with the inactivation of the 384th Bomb Group, 86-0126 was assigned to the 28th Bomb Squadron at Dyess. The aircraft was repainted and renamed HUNGRY DEVIL with new nose art. By August 31, 1996, 86-0126 was assigned to the 7th Bomb Wing and had accumulated 2,509.3 flight hours. As of August 1, 1998, the aircraft was assigned to the 28th Bomb Squadron, 7th Wing at Dyess AFB.

(Author)

86-0126, HUNGRY DEVIL, is seen here at Dyess AFB on May 12, 1997, with the blue and white checkered tail stripe of the 28th Bomb Squadron, 7th Bomb Wing. (Author)

86-0126 is seen here at McConnell AFB on August 12, 1994, with the light blue KEEPER OF THE PLAINS tail stripe and OZ tail code of the 28th Bomb Squadron, 384th Bomb Wing. (Author)

(Author)

86-0127 KANSAS LANCER

The aircraft was delivered to the 384th Bomb Wing at McConnell AFB, Kansas on February 5, 1988. The arrival nose art was also temporarily painted on 86-0127 as a backup for 86-0124 and the official arrival ceremony on February 29, 1988. It was given the name FREEDOM BIRD which was painted in an arc on both sides of the aircraft. FREEDOM BIRD had no nose art associated with the name. The name on 86-0127 was changed to IVAN'S NIGHTMARE. Nose art of a woman warrior with red-orange hair holding a sword with a circular shield at her feet was added on the left side of the aircraft. "IVAN'S" was later dropped from the name and the aircraft called simply "NIGHTMARE." As of May 31, 1992, when turned over to ACC, the aircraft had accumulated 1,444.1 flight hours. After assignment to ACC at McConnell AFB, it was assigned to the Kansas Air National Guard nicknamed KANSAS LANCER with nose art of a John Deere tractor added. By August 31, 1996, 86-0127 was assigned to the 184th Bomb Wing and had accumulated 2,713.8 flight hours. As of August 1, 1998, the aircraft was assigned to the 127th Bomb Squadron, 184th Bomb Wing, Kansas Air National Guard at McConnell AFB.

86-0127, THE KANSAS LANCER, is seen here at McConnell AFB during June 12, 1996, with the subdued gray tail stripe of the 127th Bomb Squadron, 184th Bomb Wing, Kansas ANG. (Jerry Geer)

86-0127 is seen here at McConnell AFB on March 30, 1994, in gunship gray with the light blue KEEPER OF THE PLAINS tail stripe and OZ tail code of the 28th Bomb Squadron, 384th Bomb Wing. (Author)

86-0128 PONY SOLDIER

The aircraft was delivered to the 384th Bomb Wing at McConnell AFB, Kansas on April 5, 1988. It was given the name THE HAWK which was first painted on both nose gear doors, and later was painted in an arc on both sides of the aircraft. Some of the Air Force personnel referred to the B-1 as the HAWK (an acronym for Holiday And Weekend Killer). 86-0128 was flown to Eglin AFB for follow-on climatic testing at McKinley Climatic Laboratory's climatic hangar on July 28, 1989, and returned home to McConnell AFB on September 9, 1989. On October 4, 1990, the first stage of the #1 engine failed and caused damage resulting in the engine separating from the aircraft. The aircraft landed safely at Pueblo, Colorado. On November 18, 1990 the aircraft, with only three engines installed, flew to Tinker AFB for repair and then returned to McConnell. In February 1991, the aircraft was renamed MIS BEHAVIN, and approved nose art was painted on the aircraft. As of May 31, 1992, when turned over to ACC, the aircraft had accumulated 562.4 flight hours, the fewest of any B-1 in the fleet. After assignment to ACC, and while at McConnell 86-0128 was marked as the 28th Bomb Squadron Flagship. It remained at McConnell until the 384th Bomb Group was inactivated. It moved to Ellsworth AFB during 1994. It was named PONY SOLDIER with nose art added. While at Ellsworth 86-0128 was repainted, and the nose art was removed. By August 31, 1996, 86-0128 was assigned to the 28th Bomb Wing and had accumulated 1,807.3 flight hours. As of August 1, 1998, the aircraft was assigned to the 37th Bomb Squadron, 28th Bomb Wing at Ellsworth AFB.

(Author)

86-0128, PONY SOLDIER, is seen here at Ellsworth AFB on July 24, 1995, while assigned to the 37th Bomb Squadron, 28th Bomb Wing. (Author)

86-0128 is seen here at Langley AFB on May 10, 1993, marked as the 28th Bomb Squadron Flagship. (Brian C. Rogers)

86-0129

During the summer of 1989, 86-0129, along with 86-0124 and 86-0126, took part in a deployment to Alaska and Hawaii. The aircraft was delivered to the 384th Bomb Wing at McConnell AFB, Kansas on February 26, 1988. It was given the name PEGASUS which was painted in an arc on the left side of the aircraft, with nose art of PEGASUS, the flying horse, added. As of May 31, 1992, when turned over to ACC the aircraft had accumulated 912.3 flight hours. After assignment to ACC, and while at McConnell, 86-0128 was marked as the 384th Operations Group Flagship. It remained at McConnell until the 384th Bomb Group was inactivated. It moved to Ellsworth AFB during 1994. While at Ellsworth, 86-0128 was repainted, and the nose art was removed. By August 31, 1996, 86-0129 was assigned to the 28th Bomb Wing and had accumulated 2,057.9 flight hours. As of August 1, 1998, the aircraft was assigned to the 37th Bomb Squadron, 28th Bomb Wing at Ellsworth AFB.

86-0129 is seen here at McConnell AFB on September 15, 1993, marked as the 384th Operations Group (OG) Flagship. (Author)

86-0129 is seen here at Ellsworth AFB on June 13, 1994, while assigned to the 37th Bomb Squadron, 28th Bomb Wing. (Author)

86-0130 BAD COMPANY

The aircraft was delivered to the 384th Bomb Wing at McConnell AFB, Kansas on March 3, 1988. It was given the name THE ROSE which was painted in an arc on both sides of the aircraft with no nose art. The name was later changed to BAD COMPANY which was painted on the left side of the aircraft along with nose art of a girl next to a motorcycle. As of May 31, 1992, when turned over to ACC, the aircraft had accumulated 1,252.2 flight hours. During 1994, with the inactivation of the 384th Bomb Group, 86-0126 was assigned to the 28th Bomb Squadron at Dyess. The aircraft was repainted and renamed BAD COMPANY with new nose art. By August 31, 1996, 86-0130 was assigned to the 7th Bomb Wing and had accumulated 2,665.0 flight hours. As of August 1, 1998, the aircraft was assigned to the 28th Bomb Squadron, 7th Wing at Dyess AFB.

(Author)

86-0130, BAD COMPANY, is seen here at Dyess AFB on May 12, 1997, marked with the blue and white checkered tail stripe of the 28th Bomb Squadron, 7th Bomb Wing. (Author)

86-0130 is seen here at McConnell AFB on August 21, 1992, with the light blue KEEPER OF THE PLAINS tail stripe and OZ tail code of the 28th Bomb Squadron, 384th Bomb Wing. (Author)

(Author)

86-0131

The aircraft was delivered to the 384th Bomb Wing at McConnell AFB, Kansas on March 16, 1988. It was given the name THE 8ths WONDER which was painted in an arc on both sides of the aircraft with no nose art. 86-0131 was transferred to the 319th at Grand Forks AFB on September 4, 1991 where it carried the name ULTIMATE WARRIOR with nose art of a mounted swordsman. As of May 31, 1992, when turned over to ACC, the aircraft had accumulated 1,270.7 flight hours. The aircraft was repainted, and the nose art was removed. During 1994, with the inactivation of the 384th Bomb Group, 86-0131 was assigned to the 34th Bomb Squadron at Ellsworth. By August 31, 1996, 86-0131 was assigned to the 34th Bomb Squadron and had accumulated 2,512.3 flight hours. As of August 1, 1998, the aircraft was assigned to the 34th Bomb Squadron, 366th Wing at Mountain Home AFB.

LEFT: 86-0131 is seen here at Ellsworth AFB on June 13, 1994, with the black and red tail stripe and MO tail code of the 34th Bomb Squadron, 366th Wing. (Author)

86-0131, seen here at Langley AFB on February 17, 1994, was repainted gunship gray before being reassigned to the 34th Bomb Squadron. (Brian C. Rogers)

86-0131, ULTIMATE WARRIOR, is seen here taxiing at Grand Forks AFB on July 30, 1992, marked with the blue and orange tail stripe of the 46th Bomb Squadron, 319th Bomb Wing. (Author)

86-0132 OH! HARDLUCK
28th BOMB SQUADRON FLAGSHIP

The aircraft was delivered in place at Palmdale on February 29, 1988. On August 20, 1988, 0132 was delivered to the 384th Bomb Wing at McConnell AFB, Kansas. It was given the name THE WIZARD which was painted in an arc on both sides of the aircraft with no nose art. On February 21, 1989 the aircraft was transferred to the 96th Bomb Wing at Dyess AFB, Texas and given the name OH! HARDLUCK with World War II nose art of a cat holding an old time round bomb with a burning fuse painted on both sides of the aircraft. This nose art was originally on B-17G #43-37775 of the 96th Bomb Group. 86-0132 was the first all gray B-1. As of May 31, 1992, when turned over to ACC, the aircraft had accumulated 1,706.7 flight hours. By August 31, 1996, 86-0132 was assigned to the 7th Bomb Wing and had accumulated 3,664.9 flight hours. It was marked as the 28th Bomb Squadron Flagship until replaced by 84-0056 during late 1997. As of August 1, 1998, the aircraft had retained the nickname and the nose art and was assigned to the 28th Bomb Squadron 7th Bomb Wing at Dyess AFB.

(Author)

RIGHT: 86-0132, OH! HARDLUCK, the 28th Bomb Squadron Flagship, is seen here at Nellis AFB while taking part in the USAF Golden Air Tattoo during May 1997. (Author)

86-0132, OH! HARD LUCK, is seen here taxiing at Dyess AFB on July 26, 1993, with the blue and white checkered tail stripe of the 338th CCTS, 96th Bomb Wing. (Author)

86-0132, marked as the 28th Bomb Squadron Flagship, is seen here at Dyess AFB on May 1, 1998, with the blue and white checkered tail stripe of the 28th Bomb Squadron, 7th Bomb Wing.

(Author)

86-0133

The aircraft was delivered to the 384th Bomb Wing at McConnell AFB, Kansas on April 14, 1988. It was given the name BIG BIRD. The name was later changed to THE OUTLAW which was painted in an arc on both sides of the aircraft with no nose art. The aircraft was transferred to Ellsworth AFB on October 16, 1990. As of May 31, 1992, when turned over to ACC, the aircraft had accumulated 1,080.7 flight hours. It moved to Ellsworth AFB during 1995. While at Ellsworth, 86-0133 was repainted, and the nose art was changed to BLACK HILLS BANDIT. By August 31, 1996, 86-0133 was assigned to the 28th Bomb Wing and had accumulated 2,547.0 flight hours. As of August 1, 1998, the aircraft was assigned to the 128th Bomb Squadron, 116th Bomb Wing, Georgia Air National Guard at Robins AFB.

86-0133, BLACK HILLS BANDIT, is seen here at Ellsworth AFB on June 21, 1993, marked with the blue tail stripe of the 28th Bomb Wing. (Author)

86-0133 is seen here at Robins AFB on March 21, 1997, assigned to the 128th Bomb Squadron, 116th Bomb Wing, Georgia ANG. (Brian C. Rogers)

86-0134 34th BOMB SQUADRON FLAGSHIP

The aircraft was delivered to the 384th Bomb Wing at McConnell AFB, Kansas on March 24, 1988. It was given the name GREEN HORNET, and was later changed to NIGHT MISSION. 86-0134 was renamed WILD ASS RIDE with nose art of a girl riding a bucking mule added to the left side of the aircraft. As of May 31, 1992, when turned over to ACC, the aircraft had accumulated 1,752.3 flight hours. During 1994, with the inactivation of the 384th Bomb Group, 86-0134 was assigned to the 34th Bomb Squadron at Ellsworth. The aircraft was repainted, and the nose art was removed. By August 31, 1996, 86-0134 was assigned to the 366th Wing and had accumulated 2,454.0 flight hours. As of August 1, 1998, the aircraft was marked as the 34th Bomb Squadron Flagship and assigned to the 34th Bomb Squadron, 366th Wing at Mountain Home AFB.

(Dave Pasquini)

86-0134, photographed on August 22, 1997, is marked as the 34th Bomb Squadron Flagship. (Norris Graser)

86-0134 is seen here at McConnell AFB on March 30, 1994, with the light blue KEEPER OF THE PLAINS tail stripe and OZ tail code of the 28th Bomb Squadron, 384th Bomb Wing. (Author)

(Alan Winzerling)

86-0135 DEADLY INTENTIONS

The aircraft was delivered to the 384th Bomb Wing at McConnell AFB, Kansas on April 21, 1988. It was given the name MAKE MY DAY, later changed to THE WATCHDOG. Nose of a bulldog and the U.S. Flag was added to the left side. As of May 31, 1992, when turned over to ACC, the aircraft had accumulated 1,220.9 flight hours. During 1994, with the inactivation of the 384th Bomb Group, 86-0135 was assigned to the 9th Bomb Squadron at Dyess. The aircraft was repainted and renamed DEADLY INTENTIONS with new nose art. By August 31, 1996, 86-0135 was assigned to the 7th Bomb Wing and had accumulated 2,509.8 flight hours. As of August 1, 1998, the aircraft was assigned to the 9th Bomb Squadron, 7th Wing at Dyess AFB. 86-0135 flew on both nights of B-1 operations during Operation Desert Fox. It flew on December 18, 1998 as SLAM 02, and on December 19, as SLAM 03.

86-0135, DEADLY INTENTIONS, is seen here at Dyess AFB on December 27, 1996, marked with the black tail stripe of the 9th Bomb Squadron, 7th Bomb Wing. (Alan Winzerling)

86-0135 is seen here at McConnell AFB on May 13, 1993, with the light blue KEEPER OF THE PLAINS tail stripe and OZ tail code of the 28th Bomb Squadron, 384th Bomb Wing. (Author)

86-0136

The aircraft was delivered to the 384th Bomb Wing at McConnell AFB, Kansas on April 15, 1988. It was given the name SPECIAL DELIVERY which was painted in an arc on both sides of the aircraft. Nose art of a gray stork carrying a woman, instead of a baby, in front of a gray moon was painted on the left side of the aircraft. As of May 31, 1992, when turned over to ACC, the aircraft had accumulated 1,173.3 flight hours. After assignment to ACC at McConnell AFB, it was assigned to the Kansas Air National Guard. It was repainted with no nose art. By August 31, 1996, 86-0136 was assigned to the 184th Bomb Wing and had accumulated 2,168.8 flight hours. As of August 1, 1998, the aircraft was assigned to the 127th Bomb Squadron, 184th Bomb Wing, Kansas Air National Guard at McConnell AFB.

BELOW: 86-0136 is seen here at McConnell AFB on March 23, 1998, with the subdued gray tail stripe of the 127th Bomb Squadron, 184th Bomb Wing, Kansas ANG. (Author)

BELOW CENTER: 86-0136 is seen here at McConnell AFB on August 12, 1994, still painted in the SAC three color strategic camouflage. This color scheme used flat black tail markings including the stripe of the 127th Bomb Squadron, 184th Bomb Wing, Kansas ANG. (Author)

86-0136, SPECIAL DELIVERY, is seen here at Ellsworth AFB on June 21, 1993, with the light blue KEEPER OF THE PLAINS tail stripe and OZ tail code of the 28th Bomb Squadron, 384th Bomb Wing. (Author)

(Author)

86-0137 ACE IN THE HOLE

The aircraft was delivered to the 384th Bomb Wing at McConnell AFB, Kansas on April 29, 1988. It was given the name WICHITA EXPRESS. Nose art of a charging Indian brave was designed for 0137, but was never applied to the aircraft. As of May 31, 1992, when turned over to ACC, the aircraft had accumulated 962.4 flight hours. During 1994, with the inactivation of the 384th Bomb Group, 86-0137 was assigned to the 28th Bomb Squadron at Dyess. The aircraft was repainted and renamed ACE IN HOLE with new nose art. By August 31, 1996, 86-0137 was assigned to the 7th Bomb Wing and had accumulated 2,824.1 flight hours. As of August 1, 1998, the aircraft was assigned to the 28th Bomb Squadron, 7th Wing at Dyess AFB.

86-0137 is seen here at Dyess AFB on February 20, 1995, marked with the blue and white tail stripe of the 28th Bomb Squadron, 7th Bomb Wing. (Author)

86-0137 is seen here at McConnell AFB on March 30, 1994, in gunship gray with the light blue KEEPER OF THE PLAINS tail stripe and OZ tail code of the 28th Bomb Squadron, 384th Bomb Wing. (Author)

86-0138

The aircraft was delivered in place at Palmdale on April 24, 1988. It was later delivered to the 384th Bomb Wing at McConnell AFB, Kansas. It was given the name EASY RIDER TOO with no nose art. In February 1991, 86-0138 was renamed GRAND ILLUSION II with nose art depicting a wizard approved for 0138. The art was never painted on the aircraft. As of May 31, 1992, when turned over to ACC, the aircraft had accumulated 1,298.8 flight hours. After being transferred to ACC, 86-0138 was marked as the 384th Bomb Wing Flagship. During 1994, with the inactivation of the 384th Bomb Group, 86-0138 was assigned to the 34th Bomb Squadron at Ellsworth. By August 31, 1996, 86-0138 was assigned to the 34th Bomb Squadron and had accumulated 2,715.0 flight hours. As of August 1, 1998, the aircraft was assigned to the 34th Bomb Squadron, 366th Wing at Mountain Home AFB.

(Author)

86-0138 is seen here at Langley AFB on August 6, 1993, marked as the 384th Bomb Wing Flagship with the light blue KEEPER OF THE PLAINS tail stripe and OZ tail code. (Author)

86-0138 is seen here at Ellsworth AFB on July 24, 1995, marked with the black and red tail stripe of the 34th Bomb Squadron, 366th Wing. (Author)

86-0139

The aircraft was delivered to the 384th Bomb Wing at McConnell AFB, Kansas on April 29, 1988. It was given the name GALLANT WARRIOR. In January 1990, nose art of an eagle carrying a bomb was added to the left side of the aircraft. As of May 31, 1992, when turned over to ACC, the aircraft had accumulated 1,014.2 flight hours. During 1994, with the inactivation of the 384th Bomb Group, 86-0139 was assigned to the 34th Bomb Squadron at Ellsworth. By August 31, 1996, 86-0139 was assigned to the 34th Bomb Squadron and had accumulated 2,205.8 flight hours. As of August 1, 1998, the aircraft was marked as the 34th Bomb Squadron Flagship and assigned to the 34th Bomb Squadron, 366th Wing at Mountain Home AFB.

86-0139 is seen here at Nellis AFB during September 1995, marked with the black and red tail stripe of the 34th Bomb Squadron, 366th Wing. (Both Alan Winzerling)

86-0140 LAST LANCER

The aircraft, number 100, the last B-1 built, was rolled out on January 20, 1988 at Rockwell's Palmdale Facility. It was delivered to the 384th Bomb Wing at McConnell AFB, Kansas on May 2, 1988. The main nose gear door on 86-0140 had a white shield with a red diagonal strip and the name "The Valda J", named for Major Valda J. Robbins, painted at the factory. Major Robbins was the Air Force Functional Test pilot at Palmdale who had flown the acceptance flight on all of the production B-1s. These markings were later taken off the aircraft. The aircraft was sent to Kirtland AFB during the summer and fall of 1988, and then again during the late summer of 1989 for EMP Testing. 86-0140 was later given the name PEACE WARRIOR with nose art of a warrior with sword painted on the left side of the aircraft. As of May 31, 1992, when turned over to ACC, the aircraft had accumulated 939.9 flight hours. During 1994, with the inactivation of the 384th Bomb Group, 86-0137 was assigned to the 9th Bomb Squadron at Dyess. The aircraft was repainted and renamed LAST LANCER with new nose art. By August 31, 1996, 86-0140 was assigned to the 7th Bomb Wing and had accumulated 2,390.8 flight hours. As of August 1, 1998, the aircraft was assigned to the 9th Bomb Squadron, 7th Wing at Dyess AFB.

BOTH AT RIGHT: (Author)

86-0140, LAST LANCER, is seen during on June 1997, marked with the black tail stripe of the 9th Bomb Squadron, 7th Bomb Wing. (Alan Winzerling)

86-0140 is seen here at McConnell AFB on March 30, 1994, marked with the light blue KEEPER OF THE PLAINS tail stripe and OZ tail code. (Author)

B-2 Spirit

The B-2 Spirit is a multi-role bomber and is capable of delivering both conventional and nuclear weapons. Its stealth characteristics give it the unique ability to penetrate sophisticated enemy radar defenses and attack heavily defended targets.

Northrop was awarded a $7.3 billion contract to deliver six Engineering, Manufacturing and Development (EMD) airframes and two static-test articles. The ATB's primary mission was to locate and destroy mobile Soviet ICBMs such as the SS-24 and SS-25. Development of the ATB proceeded as a "Black Program" in almost complete secrecy. The first B-2, 82-1066, rolled out at Palmdale, California on November 16, 1988.

The prime contractor responsible for overall system design and integration was Northrop Grumman's B-2 Division. Boeing, Vought Aircraft Company, Hughes Radar Systems Group and General Electric Aircraft made up the contractor team.

In 1991, procurement was cut from 132 to 75, and then even further to 15. It was later raised to 20, giving the USAF just enough aircraft for a 2 squadron wing. Finally the total was raised to 21 aircraft when Congress funded the upgrade of the prototype 82-1066 to production standards. All the other developmental aircraft will be updated to production standard. All the operational B-2s are assigned to the 509th Bomb Wing, Whiteman AFB, Missouri.

Air Vehicle-1 (AV-1) through Air Vehicle-6 (AV-6) were all developmental aircraft used for initial flight testing. When the testing was complete, the aircraft were upgraded to Block 30 standard configuration.

AV-7 to AV-16 were built to Block 10 standard (B-2A-10-NO). The Block 10 standard limited the B-2 to carriage of 16

(Northrop Grumman)

B83 free-fall nuclear weapons or MK84, 2000 pound iron bombs.

AV-17 to AV-19 were built to Block 20 standard (B-2A-20-NO). The Block 20 standard gave the aircraft the ability to carry up to 34 Cluster Bomb Units (CBUs) or 16, 2000 pound JDAMs GPS/inertially guided bombs. The GPS-Aided Targeting System (GATS) in Block 20 also allows the B-2 partial low-level terrain following capability. Block 20 aircraft can use a deployable mission planning system. This allows the mission planning system to deploy with the aircraft.

AV-20 and AV-21 were built to the Block 30 standard (B-2A-30-NO). In addition to all the Block 10 and 20 capabilities, Block 30 aircraft can carry 80 MK82 500 pound bombs, or MK62 mines, or 36 M117 750 pound bombs.

All the B-2s have been, or are being, brought to Block 30 standards.

B-2 General Characteristics
Engines: Four General Electric F-118-GE-100 engines
Thrust: 17,300 pounds each engine (7,847 kilograms)
Length: 69 feet (20.9 meters)
Height: 17 feet (5.1 meters)
Wingspan: 172 feet (52.12 meters)
Speed: High subsonic
Ceiling: 50,000 feet (15,152 meters)
Takeoff Weight (Typical): 336,500 pounds (152,635 kilograms)
Range: Intercontinental, unrefueled
Armament: Nuclear or conventional weapons
Payload: 40,000 pounds (18,144 kilograms)
Crew: Two pilots sitting side by side
Unit cost: Approximately $1.3 billion
Date Deployed: December 1993

Air Force Inventory:
Active force: 20 at Whiteman AFB and 1 at Edwards AFB

Eight B-2s can be seen in this photo of the Whiteman AFB flight line. (Northrop Grumman)

WHITEMAN AFB, MISSOURI

509th BOMB WING

The 509th Bomb Wing, one of the most famous Wings in the Air Force, traces its historical roots back to the 509th Composite Group of World War II. On August 6, 1945, the B-29 "Enola Gay" piloted by Colonel Tibbets from the 509th Bomb Group dropped the first atomic bomb and destroyed Hiroshima, Japan. On August 9, 1945, the group released an atomic bomb on Nagasaki, Japan. Within days, the Japanese sued for peace, and World War II ended.

Upon returning to the United States in late 1945, the group was based at Roswell Army Air Base, New Mexico. On November 17, 1947, SAC activated the 509th Bombardment Wing at Roswell and assigned the Group to the Wing. The Wing also received a new home in August 1958 when it moved its people and equipment to Pease AFB, New Hampshire. The Wing received its first B-52 and KC-135 in March 1966. The 509th Wing replaced its B-52s for FB-111s beginning in December 1970. The Department of Defense decided in 1988 to close Pease AFB. The 509th transferred to Whiteman AFB to become the first B-2 Stealth bomber unit, moving to Whiteman on September 30, 1990, without people and equipment. That same day also saw most of the Wing's squadrons inactivated. On June 1, 1992, when the Air Force disestablished SAC, the 509th became part of the newly created Air Combat Command. On July 20, 1993, the 509th received its first T-38 complete with a B-2-style paint job. On December 17, 1993, the first operational B-2 bomber, named "The Spirit of Missouri," touched down at Whiteman AFB. This date was also the 90th anniversary of the first powered flight by the Wright Brothers and the 49th anniversary of the original activation of the 509th Composite Group.

B-2s of the 509th Bomb Wing wear their WM (**W**hite**M**an) tail codes along with their spirit names on their main landing gear doors. The Wing's motto is Defensor Vindex (Defender and Protector)

509th Bomb Wing

(Author)

325th Bomb Squadron

The 325th Bomb Squadron was the second B-2 squadron and was activated during January 1998. The 325th Bomb Squadron was previously a B-52H Bomb Squadron assigned to the 92nd Bomb Wing at Fairchild AFB, Washington, and had been inactivated on July 1, 1994.

393rd Bomb Squadron

The 393rd Bomb Squadron was the first B-2 squadron, activating on August 27, 1993. The 393rd was part of the 509th Bombardment Group based on Tinian in the Western Pacific during World War II. It was the only squadron during World War II trained in atomic warfare, and was the Squadron which dropped both Atomic Bombs on Japan. It had been an FB-111A squadron previously assigned to the 509th Bomb Wing while based at Pease AFB, New Hampshire, and had been inactivated on September 30, 1990.

82-1066 touches down at Edwards AFB after its first flight. (Northrop Grumman)

394th Combat Training Squadron

The 394th Combat Training Squadron (CTS) traces its lineage back to the 4th Aero Squadron organized on May 5, 1917. It was redesignated as the 394th Bombardment Squadron (Heavy) on April 22, 1942, and flew B-24s as part of the 5th Bomb Group in the Pacific Theater during World War II. The Squadron was inactivated on April 29, 1946. It was activated on October 22, 1996 and assigned to the 509th Bomb Wing at Whiteman AFB on November 6, 1996. The mission of the 394th Combat Training Squadron is B-2 aircrew training.

93-1085, SPIRIT OF OKLAHOMA, and 89-0127, SPIRIT OF KANSAS, are seen here on the Whiteman flight line. (Northrop Grumman)

B-2s ASSIGNED TO 509TH BOMB WING

	A/C#	ORDER OF ARRIVAL AT WHITEMAN	DATE ARRIVED	SPIRIT
1	82-1066	20	1999	To Be Determined
2	82-1067	18	March 20, 1998	Spirit of ARIZONA
3	82-1068*	16	October 7, 1997	Spirit of NEW YORK
4	82-1069	21	1999	To Be Determined
5	82-1070	14	July 18, 1997	Spirit of OHIO
6	82-1071	19	May 23, 1998,	Spirit of MISSISSIPPI
7	88-0328	3	August 31, 1994	Spirit of TEXAS
8	88-0329	1	December 17, 1993	Spirit of MISSOURI
9	88-0330	2	August 17, 1994	Spirit of CALIFORNIA
10	88-0331	5	December 30, 1994	Spirit of S. CAROLINA
11	88-0332	4	October 30, 1994	Spirit of WASHINGTON
12	89-0127	6	February 15, 1995	Spirit of KANSAS
13	89-0128	7	June 28, 1995	Spirit of NEBRASKA
14	89-0129	8	November 17, 1995	Spirit of GEORGIA
15	90-0040	10	January 24, 1996	Spirit of ALASKA
16	90-0041	9	January 10, 1996	Spirit of HAWAII
17	92-0700	12	July 3, 1996	Spirit of FLORIDA
18	93-1085	11	May 15, 1996	Spirit of OKLAHOMA
19	93-1086	13	August 30, 1996	Spirit of KITTY HAWK
20	93-1087	15	August 5, 1997	Spirit of PENNSYLVANIA
21	93-1088	17	November 10, 1997	Spirit of LOUISIANA

* 82-1068, the Spirit of NEW YORK, is assigned to the 412th Test Wing at Edwards AFB.

MANUFACTURING

(Both - Northrop Grumman)

DE-PAINT

To keep from damaging the composite material of which the B-2 is made, Northrop developed an environmentally safe process to remove the paint from the aircraft which uses wheat starch crystals. These crystals are blown against the surfaces of the aircraft using high pressure air. (Northrop Grumman)

REPAINT

The B-2 paint hangar located at the Palmdale factory seen above is used for both de-paint and painting of the B-2 Bombers. (Northrop Grumman)

FACTORY UPGRADES

Block 10 and Block 20 aircraft were returned to the Palmdale factory for upgrades to Block 30 configuration. These two aircraft are undergoing upgrade to Block 30 configuration. (Northrop Grumman)

INSTRUMENT PANEL

(Northrop Grumman)

INDIVIDUAL B-2A SPIRITS

82-1066

The first B-2 was initially nicknamed FATAL BEAUTY and was rolled out at U.S. Air Force Plant 42, Palmdale, California on October 22, 1988; the first flight occurred on July 7, 1989, flying from Palmdale to Edwards AFB. It operated as a test aircraft assigned to the 6520th Test Squadron, 6510th Test Wing, AFMC at Edwards AFB. The unit was redesignated 420th Flight Test Squadron/412th Test Wing on October 2, 1992. 82-1066 was used for flight envelope expansion, night flights, night refuelings, and RCS-testing. It was placed in flyable storage at Palmdale in March 1993, after flying 81 test missions. It has been modified to B-2A-30-NO standard and will be used as an operational aircraft. As of August 31, 1996, 82-1066 had accumulated 344.9 test mission flight hours. At the time of publication of this book, it has not yet been given a formal "SPIRIT OF" name.

TWO BELOW: In these two photographs 82-1066 is seen on display at Palmdale on March 31, 1994. (Craig Kaston)

82-1067 SPIRIT OF ARIZONA

Aircraft number two was initially nicknamed THE SHIP FROM HELL, because if anything could go wrong on this aircraft, it did. It first flew on October 19, 1990 from Palmdale to Edwards AFB, and was operational with 6520th Test Squadron, 6510th Test Wing which changed to the 420th Flight Test Squadron, 412th Test Wing on October 2, 1992. 82-1067 was heavily instrumented and used for flutter, loads, landing gear and weapons bay door tests. By December 1993, it had flown 98 test flights for a total of 496.4 flight hours. The aircraft was later used as development aircraft with some stealth features omitted. It was later updated to Block 30 standard for operational use. As of August 31, 1996, 82-1067 had accumulated 903.3 test mission flight hours. It was named the SPIRIT OF ARIZONA at Davis-Monthan AFB, on March 20, 1998, and was the seventeenth B-2 delivered to the 509th Bomb Wing.

BELOW: (USAF photo by MSGT Dave Nolan) BOTTOM: 82-1067 is seen here in front of the hangars at Davis-Monthan AFB awaiting the ceremony naming it SPIRIT OF ARIZONA. (Norris Graser)

82-1068 SPIRIT OF NEW YORK

The third B-2A, initially nicknamed THE GHOST, first flew on June 18, 1991 from Palmdale to Edwards AFB and was operational with 6520th Test Squadron, 6510th Test Wing/420th Flight Test Squadron, 412th Test Wing. 82-1068 was the first B-2 with complete avionics and was used for radar, navigation, offensive and defensive systems testing. In February 1992 it started testing of the Northrop ZSR-63 defensive avionics system, and in November 1993 began testing of the Hughes AN/APQ-181 LPI (Low-Probability-of-Intercept) radar. It was modified to B-2A-30-NO standard. As of August 31, 1996, 82-1068 had accumulated 901.7 test mission flight hours. It was named the SPIRIT OF NEW YORK at Stewart Airport, New York, on October 7, 1997, and returned to the 412th Test Wing at Edwards AFB. It left Edwards on March 18, 1998, flying to Palmdale for Block 30 upgrades.

Seen here during October 1991, 82-1068 is on public display in its hangar at the annual Edwards AFB air show. (Mick Roth)

Seen here during October 1991, 82-1068 is on public display in its hangar at the annual Edwards AFB air show. (Paul Hart)

82-1069
B-2A number four was initially nicknamed CHRISTINE because it seemed to be possessed. It was first operational with the 6520th Test Squadron 6510th Test Wing, later 420th Flight Test Squadron, 412th Test Wing. 82-1069 was used as weapons test aircraft, releasing the first weapon, an inert MK 84 2000 pound bomb on September 3, 1992, on the Edwards Test Range. It was also used for avionics, armament, climatic, and low-observable testing. As of August 31, 1996, 82-1069 had accumulated 872.3 test mission flight hours. It was modified to B-2A-30-NO standard. At the time of publication of this book, it has not yet been given a formal "SPIRIT OF" name.

82-1069 is seen here releasing MK 82 GP 500 pound bombs. (Northrop Grumman)

82-1070 SPIRIT OF OHIO

The fifth B-2A was initially nicknamed FIRE AND ICE because it was used for climatic testing. It had art which went along with the nickname painted on the nose gear door. It first flew on October 5, 1992, and was used for low-observable comparison with AV-I. It completed six months of environmental testing at the McKinley Climatic Laboratory, Eglin AFB, Florida. As of August 31, 1996, 82-1070 had accumulated 480.0 test mission flight hours. It was modified to Block 30 standards. It was named the SPIRIT OF OHIO at Wright-Patterson AFB on July 18, 1997, and delivered to the 509th Bomb Wing at Whiteman AFB as the fourteenth operational B-2.

(Northrop Grumman)

(Northrop Grumman)

82-1071 SPIRIT OF MISSISSIPPI

The sixth B-2A was initially nicknamed BLACK WIDOW and was used for Technical Manual procedures validation at Palmdale. It first flew on February 2, 1993, flying from Palmdale to Edwards AFB on a mission lasting approximately two hours. It was operational with 420th Flight Test Squadron/412th Test Wing. As of August 31, 1996, 82-1071 had accumulated 571.7 test mission flight hours. It was modified to B-2A-30-NO standard and delivered to the 509th Bomb Wing. It was named SPIRIT OF MISSISSIPPI on May 23, 1998, at Thompson Field Air National Guard Base, Jackson, Mississippi. It was delivered to the 509th Bomb Wing at Whiteman AFB, Missouri as the unit's eighteenth operational B-2.

The USAF's two stealth combat aircraft, B-2 82-1071 and an F-117 from the 412th Test Wing are seen here over the southern Sierra Nevada Mountains. (Northrop Grumman)

88-0328 SPIRIT OF TEXAS

The number seven B-2 was built as a B-2A-10-NO, modified to B-2A-20-NO configuration, further modified to B-2A-30-NO configuration. It was initially nicknamed THE PIRATE SHIP. It was leased back to Northrop for electromagnetic compatibility (EMC) and emission security testing, which was completed in August 1993. It was named the SPIRIT OF TEXAS at Naval Air Station Dallas, Texas on August 1994 and delivered to the 509th Bomb Wing at Whiteman AFB as the third operational B-2. As of August 31, 1996, 88-0328 had accumulated 492.0 flight hours.

(All - Keith Snyder)

88-0329 SPIRIT OF MISSOURI

The number eight B-2 was built as a B-2A-10-NO, modified to B-2A-20-NO configuration, further modified to B-2A-30-NO configuration. 88-0329 was the first B-2 delivered to the 509th Bomb Wing. It was named SPIRIT OF MISSOURI at its delivery ceremony held at Whiteman AFB on December 17, 1993. It flew to the Paris Air Show on June 10 and 11, 1995. As of August 31, 1996, 88-0329 had accumulated 504.4 flight hours.

Bombers of the 509th Bomb Wing, the B-29, B-52, FB-111A, and the B-2, are seen here posed on the Whiteman AFB flight line. (Northrop Grumman)

(Northrop Grumman)

88-0330 SPIRIT OF CALIFORNIA

88-0330 was the ninth built, and was built as a B-2A-10-NO, modified to B-2A-20-NO configuration, further modified to B-2A-30-NO configuration. It was named the SPIRIT OF CALIFORNIA at Palmdale on August 17, 1994, and was the second B-2 delivered to 509th Bomb Wing. As of August 31, 1996, 88-0330 had accumulated 609.9 flight hours.

(All - Craig Kaston)

88-0331 SPIRIT OF SOUTH CAROLINA

Air Vehicle ten was built as a B-2A-10-NO, modified to B-2A-20-NO configuration, further modified to B-2A-30-NO configuration. It was named SPIRIT OF SOUTH CAROLINA on December 30, 1994 at Shaw AFB, becoming the 509th Bomb Wing's fifth operational B-2. As of August 31, 1996, 88-0331 had accumulated 516.9 flight hours.

BELOW: (Northrop Grumman)

Because of the delicate nature of the radar absorbing materials and the composite materials of which the B-2 are made, the aircraft at Whiteman AFB are not kept outside on the flightline, but rather have their own hangars. 88-0331 is seen here in its parking hangar. (Northrop Grumman)

88-0332 SPIRIT OF WASHINGTON

The 11th B-2 was built as a B-2A-10-NO, modified to B-2A-20-NO configuration, further modified to B-2A-30-NO configuration. It was named SPIRIT OF WASHINGTON on October 30, 1994, at Boeing Field, Seattle, Washington, becoming the 509th Bomb Wing's fourth operational B-2. As of August 31, 1996, 88-0332 had accumulated 544.1 flight hours.

RIGHT: (Northrop Grumman)
BELOW: (R.E.F. Jones)

89-0127 SPIRIT OF KANSAS

89-0127, the twelfth B-2, was built as a B-2A-10-NO, modified to B-2A-20-NO configuration, further modified to B-2A-30-NO configuration. It was named SPIRIT OF KANSAS on February 17, 1995 at McConnell AFB, becoming the 509th Bomb Wing's sixth operational B-2. As of August 31, 1996, 89-0127 had accumulated 239.5 flight hours. It stopped over at RAF Fairford on July 19, 1997.

TWO BELOW: (Author)
BOTTOM: (Northrop Grumman)

89-0128 SPIRIT OF NEBRASKA

The thirteenth B-2 was built as a B-2A-10-NO, modified to B-2A-20-NO configuration, further modified to B-2A-30-NO configuration. It arrived at Whiteman on June 28, 1995, and was named SPIRIT OF NEBRASKA on September 3, 1995 at Offutt AFB, becoming the 509th Bomb Wing's seventh operationa B-2. As of August 31, 1996, 89-0128 had accumulated 341.8 flight hours.

(Author)

89-0129 SPIRIT OF GEORGIA

The 89-0129 was built as a B-2A-10-NO, modified to B-2A-20-NO configuration, further modified to B-2A-30-NO configuration. It was named SPIRIT OF GEORGIA on November 14, 1995 at Robins AFB, becoming the 509th Bomb Wing's eighth operational B-2.

(Both - Northrop Grumman)

90-0040 SPIRIT OF ALASKA

B-2 90-0040 was built as a B-2A-10-NO, modified to B-2A-20-NO configuration, further modified to B-2A-30-NO configuration. It was named SPIRIT OF ALASKA on January 24, 1996 at Elmendorf AFB, becoming the 509th Bomb Wing's tenth operational B-2. As of August 31, 1996, 90-0040 had accumulated 196.4 flight hours.

(Northrop Grumman)

90-0041 SPIRIT OF HAWAII

The number sixteen B-2 was built as a B-2A-10-NO, modified to B-2A-20-NO configuration, further modified to B-2A-30-NO configuration. It was named SPIRIT OF HAWAII on January 10, 1996 at Hickam AFB, Hawaii, becoming the 509th Bomb Wing's ninth operational B-2. As of August 31, 1996, 90-0041 had accumulated 290.6 flight hours.

(Northrop Grumman)

92-0700 SPIRIT OF FLORIDA

Air vehicle number seventeen, the only "1992 model", was built as a B-2A-10-NO, modified to B-2A-20-NO configuration, further modified to B-2A-30-NO configuration. It was named SPIRIT OF FLORIDA on July 3, 1996 at MacDill AFB, becoming the 509th Bomb Wing's twelfth operational B-2. As of August 31, 1996, 92-0700 had accumulated 115.2 flight hours.

RIGHT: (USAF) BELOW: (Northrop Grumman)

93-1085 SPIRIT OF OKLAHOMA

B-2 number eighteen was built as a B-2A-20-NO and modified to B-2A-30-NO configuration. It was named SPIRIT OF OKLAHOMA on May 15, 1996 at Tinker AFB, becoming the 509th Bomb Wing's eleventh operational B-2. As of August 31, 1996, 93-1085 had accumulated 99.9 flight hours.

(Both - Northrop Grumman)

93-1086 SPIRIT OF KITTY HAWK

93-1086 was initially built as a B-2A-20-NO and modified to B-2A-30-NO configuration. It was named SPIRIT OF KITTY HAWK on August 30, 1996 at Seymour-Johnson AFB, becoming the 509th Bomb Wing's thirteenth operational B-2. As of August 31, 1996, 93-1086 had accumulated 26.2 flight hours.

(Northrop Grumman)

93-1087 SPIRIT OF PENNSYLVANIA

The twentieth B-2 built, the first built as a B-2A-30-NO, was named SPIRIT OF PENNSYLVANIA on August 5, 1997 at Willow Grove Air Reserve Base, becoming the 509th Bomb Wing's fifteenth operational B-2.

(All - Don Spering AIR)

93-1088 SPIRIT OF LOUISIANA

The twenty-first and last B-2 was built as a B-2A-30-NO. It was named SPIRIT OF LOUISIANA on November 10, 1997 at Barksdale AFB, becoming the 509th Bomb Wing's sixteenth operational B-2. It deployed to Andersen AFB, Guam during April, 1998, as part of the B-2's first operational overseas deployment.

(Northrop Grumman)

(Both: Brian C. Rogers)

Bomber Weapons

SUSPENSION SYSTEMS

B-52 WEAPON SUSPENSION AND RELEASE SYSTEMS

BOMB BAY SUSPENSION

The bomb bay of the B-52 occupies almost the entire center section of the fuselage, between the two sets of main landing gear, and is 28 feet long and 6 feet wide. It is enclosed by double-panel doors. The inner panels of the doors are made up of three interconnected sections which open for weapon release, with both the inner and outer panels opening on the ground for weapon loading.

Bomb Racks

The B-52 can carry two four-weapon cluster racks or three nine-weapon clip-in racks mounted internally. Nuclear weapons are no longer carried on the racks.

The bomb bay of the B-52 takes up the fuselage space just aft of the wing and between the forward and aft main landing gear. (Author's collection)

Common Strategic Rotary Launcher

The Common Strategic Rotary Launcher (CSRL) is an eight station rotary launcher and presently is the only internal nuclear capable suspension system carried by the B-52. The CSRL was developed as part of Cruise Missile Integration. The B-52's previous rotary launcher had been developed to carry the Short Range Attack Missile (SRAM) and was only 15 feet (180 inches) long. The new cruise missile being developed for the B-52 was over twenty feet long, and as a result a longer launcher had to be developed. The CSRL, over 25 feet long, gave the B-52 the capability to carry internal AGM-86Bs. The launcher is also capable of carrying B-53, B-61, and B-83 gravity nuclear bombs

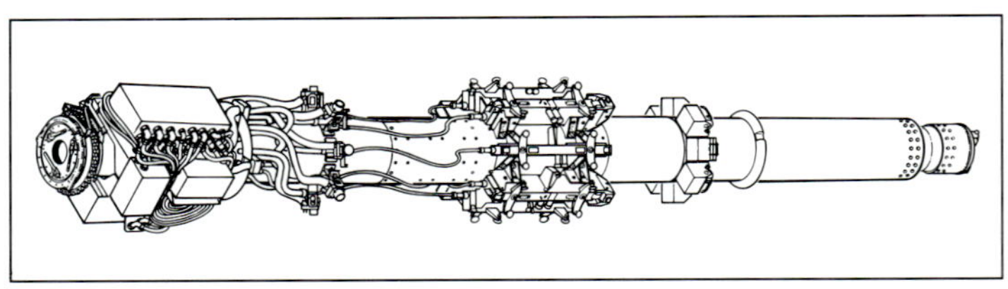

Common Strategic Rotary Launcher (CSRL)

The CSRL installation program began during 1988. The modification added supports for the yokes which suspended and rotated the launcher within the bomb bay. Hydraulic lines, and electric and data bus cabling was added. The CSRL modifications were accomplished at the B-52 depots (Tinker AFB, Oklahoma and Kelly AFB, Texas). Eighty-two B-52Hs were modified, fourteen were not. Two CSRL capable B-52Hs (60-0026 and 60-0040) have been lost, leaving 80 still in the fleet.

The CSRL is capable of carrying eight AGM-86B ALCMs, eight B61-7 nuclear bombs, eight B83-0 or B83-1 nuclear bombs, or eight AGM-86C CALCMs. The avionics software does not allow mixing of the weapon types on the launcher.

B-52 EXTERNAL PYLONS

Three types of external pylons can be fitted underneath the wings of the B-52. The tallest pylon, originally used for carrying the AGM-28 Hound Dog cruise missile, is known as the AGM-28 pylon. The shorter one, known as a "stub" pylon, was designed for conventional weapons carriage. The pylons can be fitted with a nine station Heavy Stores Adapter Beam (HSAB) for weapons up to 3,000 pounds each, or an I-beam Multiple Ejector Rack (MER) adapter, which can carry two six-station MERs with up to 12 weapons on each pylon.

The third pylon type is the cruise missile pylon which can carry six ALCM or CALCM on the cruise missile underwing pylons. It can also carry six AGM-129 Advanced Cruise Missiles (ACM).

This photo shows a MER adapter Beam attached to an AGM-28 pylon under the right wing of a B-52H. (Author)

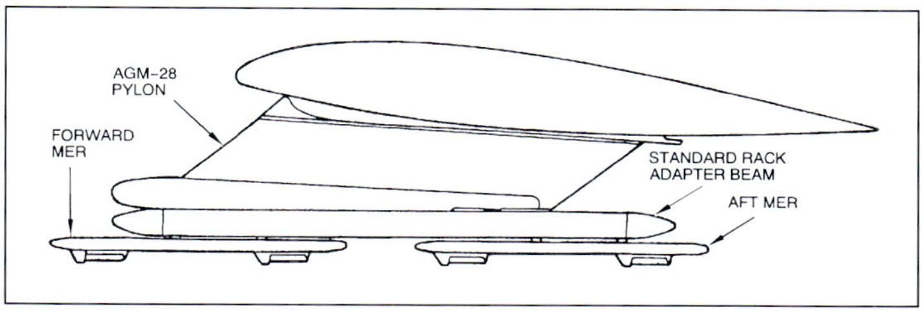

Multiple Ejector Rack (MER) Beam

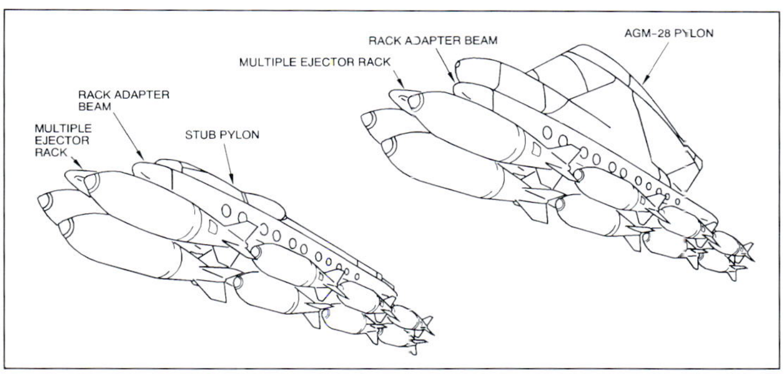

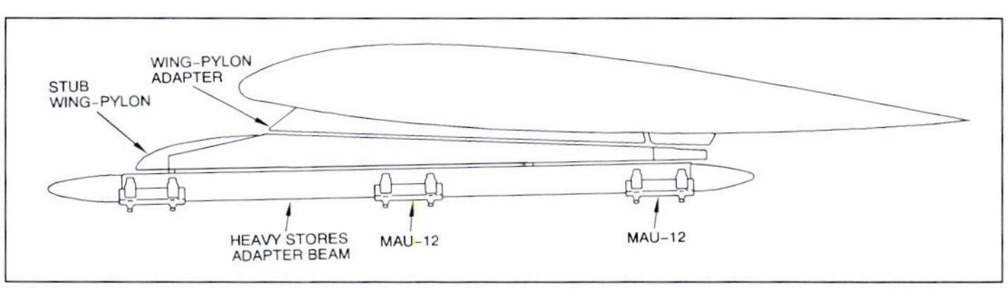

Heavy Stores Adapter Beam (HSAB)

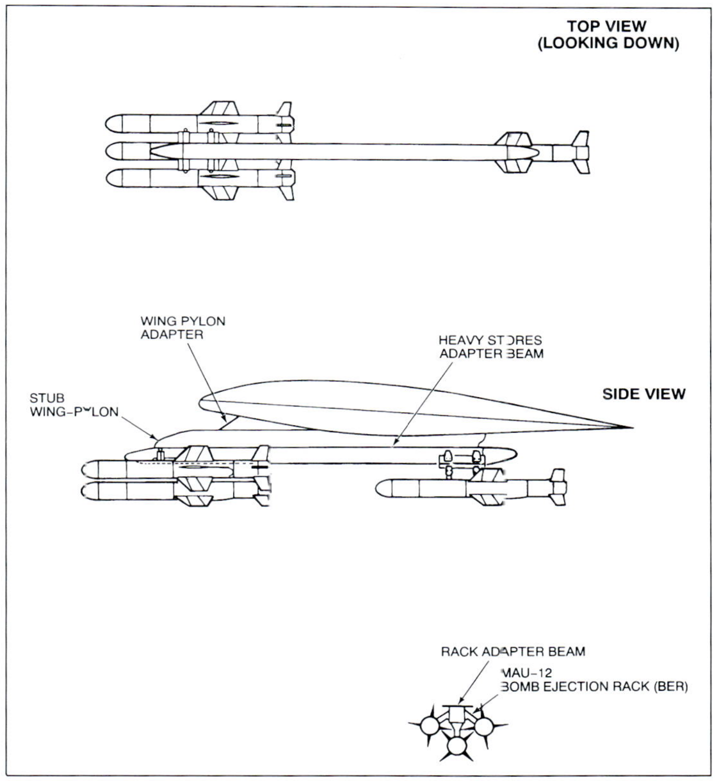

GM-84 Harpoon Carriage

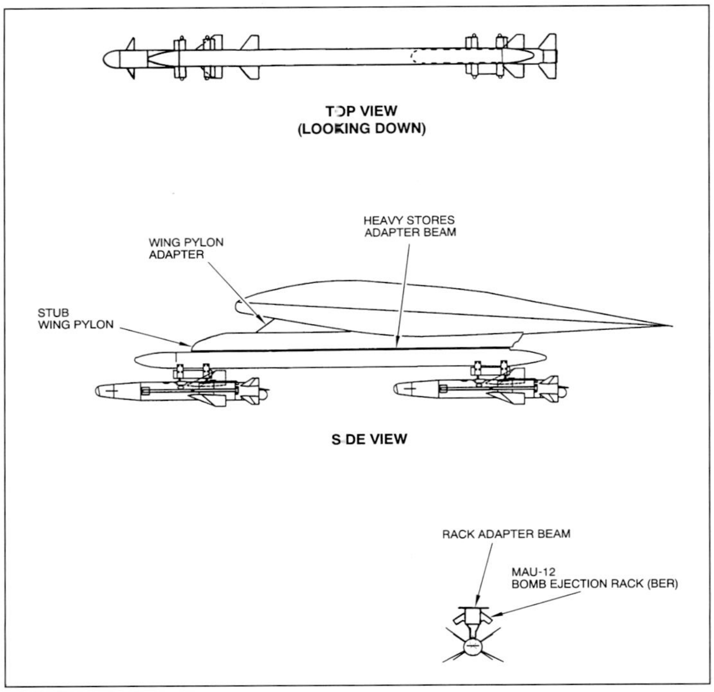

AGM-142 Have Nap Carriage

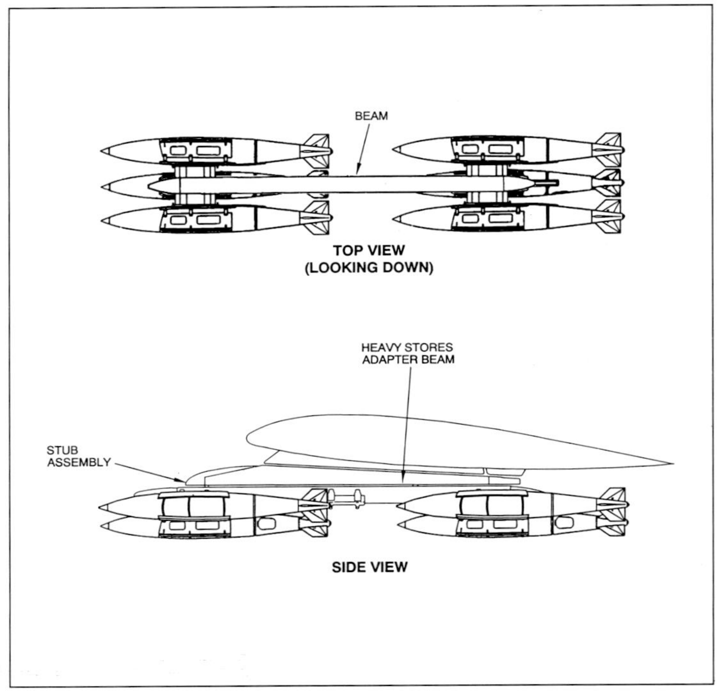

GBU-31 JDAM Carriage

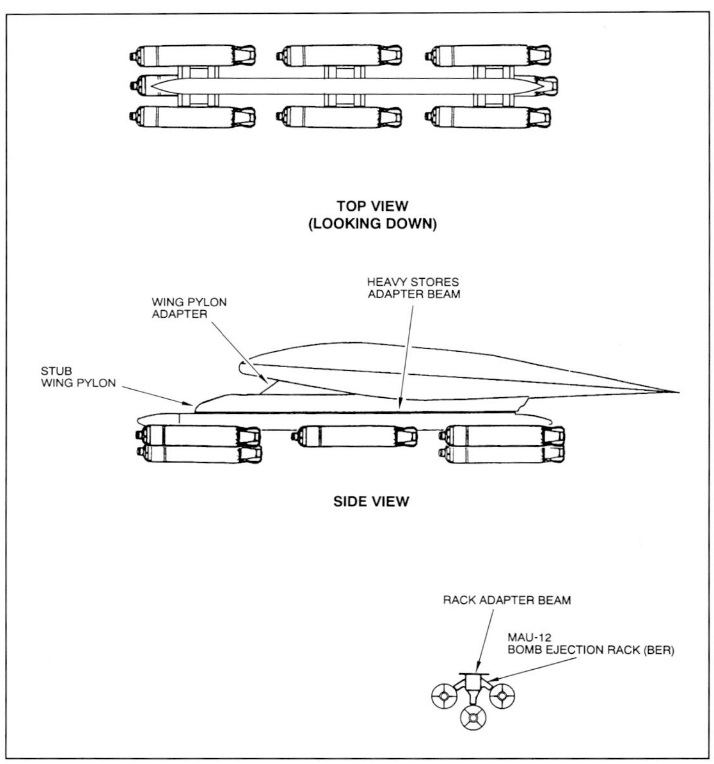

WCMD Carriage

Cruise Missile Pylon

The third pylon type is the cruise missile pylon which can carry six ALCM or CALCM on the cruise missile underwing pylons. It can also carry six AGM-129 Advanced Cruise Missiles (ACM).

A cruise missile pylon is visible under the left wing of 61-036 seen here in 5th Bomb Wing, Minot AFB markings. (USAF photo)

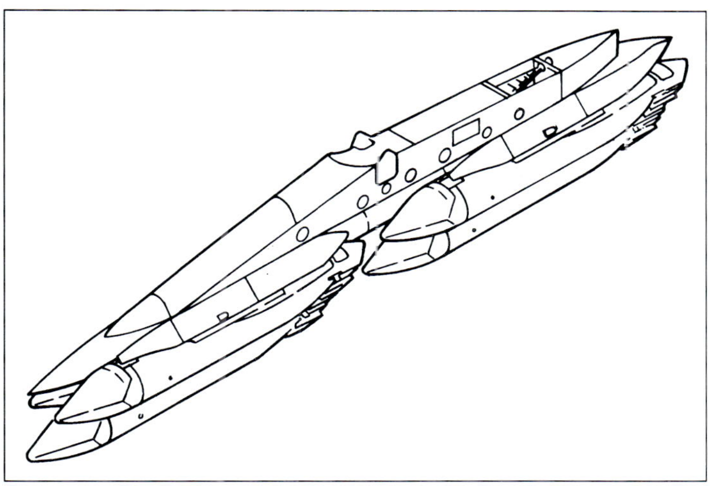

Cruise Missile Pylon with AGM-129As.

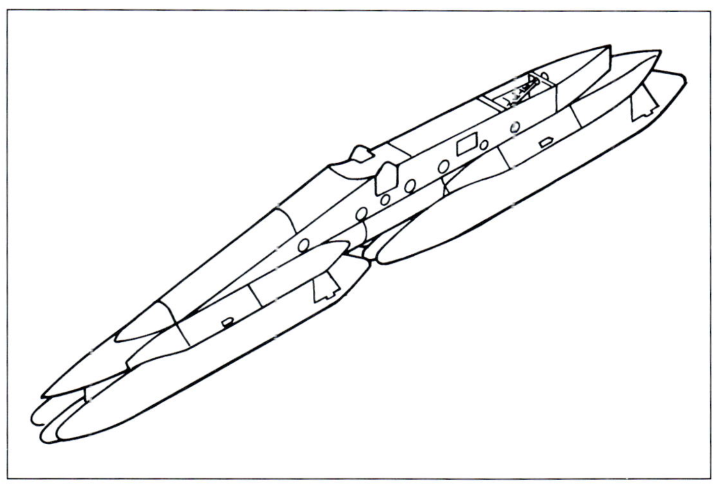

Cruise Missile Pylon with AGM-86.

LEFT: The pylon at wing station 559 between the engine pods was originally used for AN/ALE-25 forward firing chaff rockets. It has recently been used for the ACMI pod used on the Nellis AFB ranges. Testing of the pylon for carriage of the Have Nap Data Link Pod has also been accomplished. (Author) RIGHT: Power to the ACMI pod, seen here, is controlled from the Navigators station. (Brian C. Rogers)

B-52H Weapons

WEAPON TYPE	NO. LOADED CLUSTER RACKS (THREE RACKS)	NO. LOADED CLIP-IN (TWO RACKS)
	INTERNAL	
NONNUCLEAR		
CBU		
CBU-52B/B	27	
CBU-58/B	27	
CBU-71/B	27	
CBU-87	6	
CBU-89/B	6	
BOMBS		
MK82	27	
MK82 AIR	27	
MK84		8
M117	27	
M117R	27	
M117AIR	27	
M117D	27	
BDU-48	9	
M129/M129E-1 Leaflet	18	
MJU-1/B	18	
MINES		
MK36 (DST)	27	
MK52	12	
DST (M117) MK59	27	
MK62 (QS)	27	
MK55		8
MK56 Unfinned		8
MK60		8
MK65 (QS)		8
NUCLEAR		
	CSRL	
B61-7	8	
B83-0/B83-1	8	
BDU-38/B Practice	8	
AGM-86B ALCM	8	
NONNUCLEAR		
AGM-86C (CONV ALCM)	8	

B-52H Weapons (Cont)

WEAPON TYPE	NO. LOADED (MER)	NO. LOADED (HSAB)
	EXTERNAL	
NONNUCLEAR		
BOMBS		
MK82	24	18
MK82 AIR	24	18
MK84		10
MC-1	24	18
MC-1 HD	24	18
M117	24	18
M117R	24	18
M117AIR	24	18
M117D	24	18
BDU-48	8 (max)	
M129/M129E-1 Leaflet	24	18
MJU-1/B	24	18
CBU		
MK20 (Rockeye)	24	18
CBU-52B/B	24	18
CBU-58/B	24	18
CBU-71/B	24	18
CBU-87	22	18
CBU-89/B	24	18
LGB		
GBU-10/B, A/B PW I		8
GBU-10C/B, D/B, E/B PW II		10
GBU-10G/B, H/B, J/B PW II		10
GBU-12B/B, C/B, D/B PW II		10
JDAM		
GBU-31(V)1/B		12
GBU-31(V)3/B		12
WCMD		
CBU-103		16
CBU-104		16
CBU-105		16

B-52H Weapons (Cont)

WEAPON TYPE	NO. LOADED (MER)	NO. LOADED (HSAB)
	EXTERNAL (Cont)	
MINES		
MK36 (DST)	24	18
MK-40 (DST) (w/MAU-91 Fin)		18
MK-40 (DST) (w/MK-12 Mod 0 Tail)		18
MK52		18
MK55		12
MK56 Unfinned		12
MK56 Finned		12
DST (M117) MK59	24	18
MK60		10
MK62 (QS)	24	18
MK63 (QS) (w/MAU-91 Fin)		18
MK63 (QS) (w/MK-12 Mod 0 Tail)		18
MK65 (QS)		10
MISSILES		
AGM-84 HARPOON		8
AGM-142 HAVE NAP (If Data Link Pod is carried only three missiles can be carried)		4
NUCLEAR	**CRUISE MISSILE PYLON**	
AGM-86B ALCM	12	
AGM-129 ACM	12	
NONNUCLEAR		
AGM-86C (CONV. ALCM)	12	

B-1 WEAPON SUSPENSION AND RELEASE SYSTEMS

The B-1 has three stores bays (weapons bays), two between the nose landing gear and the main landing gear, and one aft of the main landing gear. Each stores bay can hold one rotary launcher, or one of two types of conventional bomb modules, or an auxiliary fuel tank. Present avionics software loads allow carriage of only all bomb modules or all rotary launchers at the same time. Future avionics software changes will allow the rotary launchers and bomb modules to be mixed in different bays on the same aircraft. All the weapons suspension equipment is attached to an attach point on the forward and aft bulkheads of the bay.

CONVENTIONAL ROTARY LAUNCHER

The Conventional Rotary Launcher (CRL) can carry up to eight 2000 pound class nonnuclear weapons. The CRLs are carried internally and can be mounted in each of the three stores bays. The CRL is attached to the forward and aft stores bay bulkheads in the same manner as all other B-1 weapon suspension equipment.

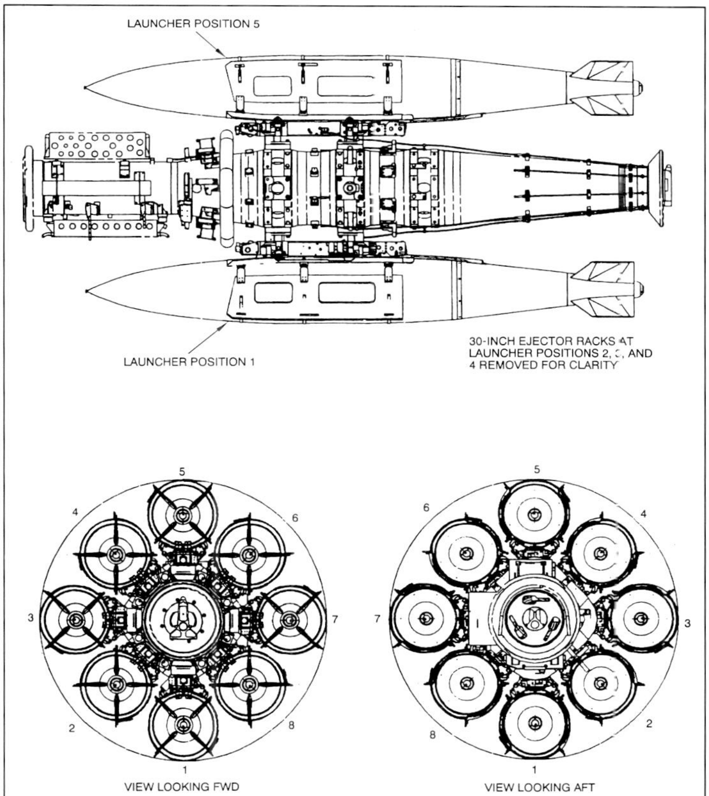

GBU-31(V)1/B CONFIGURATION

B-1 Conventional Rotary Launcher (CRL)

CONVENTIONAL BOMB MODULES

Two types of Conventional Bomb Modules (CBM), the 28 carry CBM and the 10 carry CBM, can be carried in all three store bays on the B-1. The 28 carry CBM and the 10 carry CBM may be mixed on a single aircraft.

28 Carry Conventional Bomb Module

The 28 carry CBM is used for individual release of 500 pound class nonnuclear weapons. The module may be installed in any of the three stores bays using the same attach points as the rotary launchers. When the module is installed, the rotary launcher drive system is mechanically locked and electrically disabled.

10 Carry Conventional Bomb Module

The CBM is used for individual release of 1000 pound class nonnuclear weapons. The module may be installed in any of the three stores bays using the same attach points as the rotary launchers. When the module is installed, the rotary launcher drive system is mechanically locked and electrically disabled.

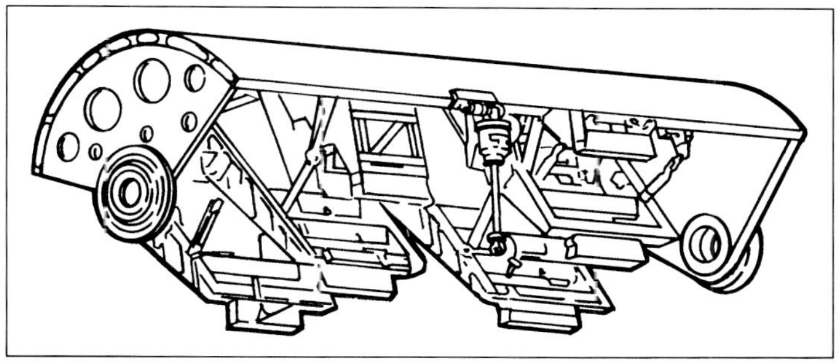

28 Carry Bomb Module.

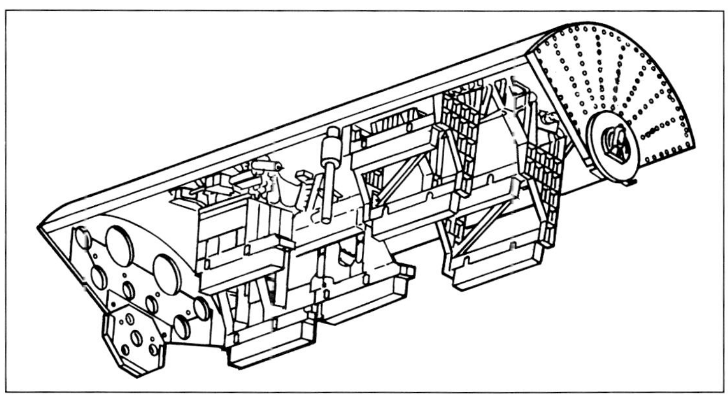

10 Carry Bomb Module.

B-1 Weapons

WEAPON TYPE	10 CARRY MODULE		28 CARRY MODULE		ROTARY LAUNCHER		MAXIMUM TOTAL
	QUANTITY	BAYS	QUANTITY	BAYS	QUANTITY	BAYS	
NONNUCLEAR							
BOMBS							
MK82 LD	7	Fwd, Mid, Aft	28	Fwd, Mid, Aft	–	–	84
MK82 AIR	7	Fwd, Mid, Aft	28	Fwd, Mid, Aft	–	–	84
MK84 GP/LD	–	None	–	None	8	Fwd, Mid, Aft	24
BDU33C	0	None	10	Mid, Aft	–	–	30
BDU33D	5	Mid, Aft	8	Mid, Aft	–	–	24
CBU							
CBU-87	10	Fwd, Mid, Aft	–	None	–	None	30
CBU-89/B	10	Fwd, Mid, Aft	–	None	–	None	30
CBU-97/B	10	Fwd, Mid, Aft	–	None	–	None	30
MINES							
MK62 QS	7	Fwd, Mid, Aft	28	Fwd, Mid, Aft	–	–	84
JDAM							
GBU-31(V)1/B	–	None	–	None	8	Fwd, Mid, Aft	24
GBU-31(V)3/B	–	None	–	None	8	Fwd, Mid, Aft	24
WCMD (Block E Software)							
CBU-103(V)2	10	Fwd, Mid, Aft	–	None	–	None	30
CBU-104(V)2	10	Fwd, Mid, Aft	–	None	–	None	30
CBU-105(V)2	10	Fwd, Mid, Aft	–	None	–	None	30
JSOW							
AGM-154	–	None	–	None	8	Fwd, Mid, Aft	24
NUCLEAR							
B-61-7	–	None	–	None	8	Fwd, Mid, Aft	24
B-83	–	None	–	None	8	Fwd, Mid, Aft	24

B-2 WEAPON SUSPENSION AND RELEASE SYSTEMS

The B-2 has two side-by-side weapons bays. Each weapons bay can hold one rotary launcher or two bomb racks.

ROTARY LAUNCHER

Each rotary launcher holds up to eight weapons. Weapon ejector racks can be installed in staggered positions forward and aft to allow weapon fin clearance. The launcher is supported on forward and aft fittings on the weapons bay bulkheads. The forward fitting provides connections for cooling air and mechanical rotation. The aft fitting provides support and electrical connection.

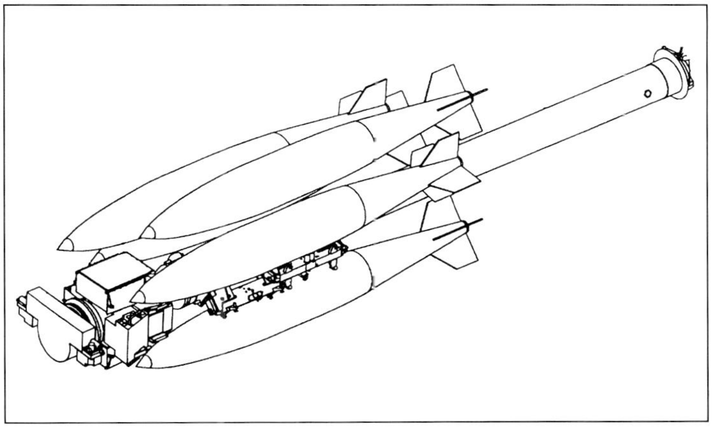

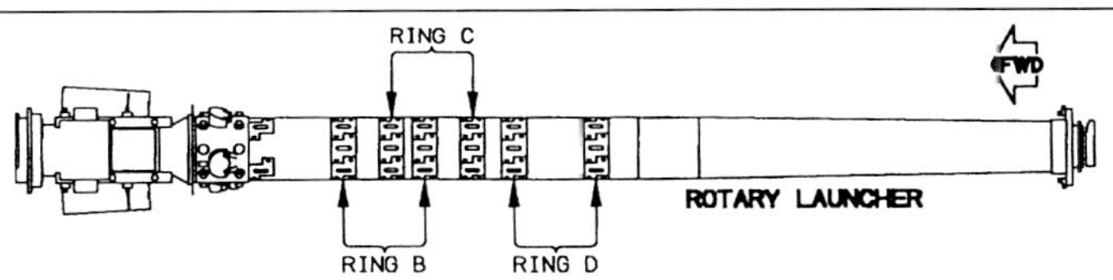

RING ASSIGNMENT — NONNUCLEAR
(NUMBER OF WEAPONS PER RING)

	RING B	RING C	RING D
1		MK 84 LDGP (0 TO 4)	MK 84 LDGP (0 TO 4)
2		GBU-31(V) 1/B (0 TO 4)	GBU-31(V) 1/B (0 TO 4)
3		GBU-31(V) 3/B (0 TO 4)	GBU-31(V) 3/B (0 TO 4)
4		GBU-36/B (0 TO 4)	GBU-36/B (0 TO 4)
5		GBU-37/B (0 TO 4)	

- EACH ROW SHOWS AN AUTHORIZED WEAPON LOAD.
- MAXIMUM WEAPON LOAD FOR EACH LAUNCHER IS EIGHT.

RING ASSIGNMENT — NUCLEAR
(NUMBER OF WEAPONS PER RING)

	RING B	RING C	RING D
1	B83-0/1 (0 TO 4) B61-7 (0 TO 4)		B83-0/1 (0 TO 4) B61-7 (0 TO 4)
2	B83-0/1 (0 TO 4)		B61-11 (0 TO 4) B83-0/1 (0 TO 4)

- EACH ROW SHOWS AN AUTHORIZED WEAPON LOAD.
- MAXIMUM WEAPON LOAD FOR EACH RING IS FOUR AND FOR EACH LAUNCHER IS EIGHT.

BOMB RACKS

Two bomb rack types are used with the B-2. One carries twenty 500 pound class weapons in four rows of five, the other carries nine 1000 pound weapons in three rows of three. Two bomb racks can be installed in each weapons bay.

Bomb Rack Type A

The bomb rack A configuration can carry up to twenty bombs in four columns of five bombs each.

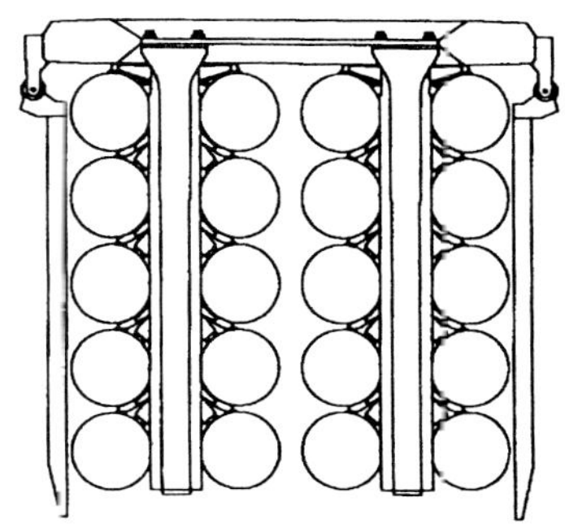

WEAPON	WEAPONS PER COLUMN			
	OUTBOARD	CENTER OUTBOARD	CENTER INBOARD	INBOARD
MK 82 LDGP	5	5	5	5
MK 82 AIR LOW DRAG HIGH DRAG	 5 5	 5 5	 5 5	 5 5
MK 62 QS	5	5	5	5

Bomb Rack Type B

The bomb rack B configuration can carry up to nine bombs in three columns of three bombs each. Cluster bomb units are not loaded on the bottom inboard station of either forward bomb rack.

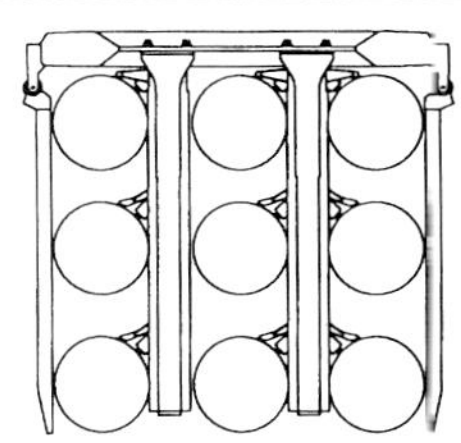

WEAPON	WEAPONS PER COLUMN		
	OUTBOARD	CENTER	INBOARD
CBU-87/B FORWARD RACKS AFT RACKS	 3 3	 3 3	 2* 3
CBU-89/B FORWARD RACKS AFT RACKS	 3 3	 3 3	 2* 3
CBU-97/B FORWARD RACKS AFT RACKS	 3 3	 3 3	 2* 3
M117 GP FORWARD RACKS AFT RACKS	 3 3	 3 3	 3 3
M117R FORWARD RACKS AFT RACKS	 3 3	 3 3	 3 3

* EACH BOTTOM INBOARD STATION MUST BE EMPTY.

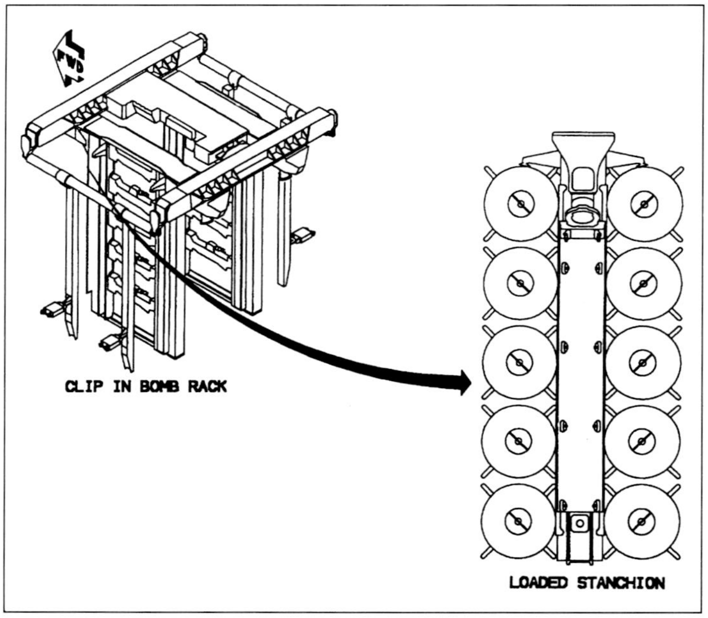

B-2 Weapons

WEAPON TYPE	BOMB RACK A	BOMB RACK B	ROTARY LAUNCHER	MAXIMUM TOTAL
NONNUCLEAR				
BOMBS				
MK82 LDGP	20 Each Rack	–	–	80
MK82 AIR	20 Each Rack	–	–	80
M117 GP	–	9	–	36
M117 R	–	9	–	36
MK84 LDGP	–	–	8	16
GPS AIDED BOMBS				
GBU-36/B	–	–	8	16
GBU 37/B	–	–	4	8
CBU				
CBU-87/B	–	8 Fwd, 9 Aft	–	34
CBU-89/B	–	8 Fwd, 9 Aft	–	34
CBU-97/B	–	8 Fwd, 9 Aft	–	34
MINES				
MK62 QS	20 Each Rack	–	–	80
JDAM				
GBU-31(V)1/B	–	–	8	16
GBU-31(V)3/B	–	–	8	16
NUCLEAR				
B-61-7	–	–	8	16
B-61-11	–	–	8	16
B-83	–	–	8	16

WEAPONS

AIR TO GROUND MISSILES

AGM-84D	Harpoon
AGM-86B	Air Launched Cruise Missile (ALCM)
AGM-86C	Conventional Air Launched Cruise Missile (CALCM)
AGM-129	Advanced Cruise Missile (ACM)
AGM-142	Have Nap
AGM-154	Joint Stand Off Weapon (JSOW)
AGM-158	Joint Air to Surface Standoff Missile (JASSM)

NUCLEAR GRAVITY BOMBS

B-61
B-83

GUIDED WEAPONS

GBU-31/	MK-84 JDAM and BLU 109 JDAM
GBU-36/B	MK-84 GPS Aided Targeting System - GPS Aided Munition (GATS-GAM)
GBU-37/B	GAM-113 GPS Aided Targeting System - GPS Aided Munition (GATS-GAM)

LASER GUIDED BOMBS

GBU-10/B, A/B	MK-84 Paveway I
GBU-10C/B, D/B, E/B	MK-84 Paveway II
GBU-10G/B, H/B, J/B	BLU-109 Paveway II
GBU-12B/B, C/B, D/C	MK-82 Paveway II

CLUSTER BOMB UNITS

CBU-52/B	Cluster Bomb (No Longer Operational)
CBU-58/B	Cluster Bomb (No Longer Operational)
CBU-71/B	Cluster Bomb (No Longer Operational)
CBU-87/B	Combined Effects Munition (CEM)
CBU-89/B	GATOR
CBU-97/B	Sensor Fuzed Weapon (SFW)
CBU-103	CEM Wind Corrected Munitions Dispenser (WCMD)
CBU-104	GATOR WCMD
CBU-105	SFW WCMD

GENERAL PURPOSE BOMBS

MK 84 SERIES	2000 Pound General Purpose Bomb
MK 82 SERIES	500 Pound General Purpose Bomb
M117 SERIES	750 Pound General Purpose Bomb

MINES

MK 56	Moored Mine
MK 60	CAPTOR Mine
MK 59 QS	M-117 DST Quick Strike Mine
MK 62 QS	500 Pound Quick Strike Mine
MK 63 QS	1000 Pound Quick Strike Mine
MK 65 QS	2000 Pound Quick Strike Mine

AGM-84 HARPOON ANTI-SHIPPING MISSILE (B-52)

Originally developed for the U.S. Navy as an anti-ship missile, the AGM-84D Harpoon also has been added to the arsenal of the B-52. The AGM-84D Harpoon is an all-weather, over-the-horizon, anti-ship missile system produced by McDonnell Douglas. Its low-level, sea-skimming cruise trajectory, active radar guidance and warhead design assure high survivability and effectiveness. B-52 bombers can carry up to eight missiles.

The Harpoon was fully integrated into a fully operational B-52G from Mather AFB, California, in March 1983. Three successful live launches at the Naval Air Warfare Center, Point Mugu, California, led to the modification of a total of 30 B-52Gs with Harpoon launch control equipment, enough to provide two squadrons of Harpoon-capable B-52Gs. The 42nd Bombardment Wing, Loring AFB, Maine, and the 43rd Bombardment Wing, Andersen AFB, Guam, were first tasked to perform the Harpoon Mission.

After Loring AFB closed and the retirement of the last B-52G at Castle AFB, California, the Harpoon Mission was moved to the B-52Hs of the 2nd Bomb Wing at Barksdale AFB, Louisiana. Four B-52H models were rapidly modified as part of the Rapid Eight as an interim measure to accept B-52G Harpoon launch control equipment pending B-52H fleet modification. During 1997, a new Stores Management Overlay (SMO) for the Offensive Avionics System allowed all ICSMS B-52H airframes to carry Harpoon missiles. Both the 5th Bomb Wing at Minot AFB, North Dakota, and the 2nd Bomb Wing at Barksdale, have Harpoon capable B-52s.

The four Harpoon B-52H aircraft of the Rapid Eight were 60-0013, 60-0038, 61-0013, and 60-0019.

Primary Function: Air-to-surface anti-ship missile
Engine: Teledyne Turbojet
Thrust: 660 pounds (approximately 594 kilograms)
Length: 12 feet, 7 inches (3.79 meters)
Wing Span: 3 feet (91.44 centimeters)
Diameter: 13.5 inches (34.29 centimeters)
Weight: 1,145 pounds (515.25 kilograms)
Range: Over the horizon
Speed: High subsonic
Guidance System: Sea-skimming cruise monitored by radar altimeter, active radar terminal homing Warheads: Penetration high-explosive blast (488 pounds)

This weapon is an ATM-84A training missile which does not have a warhead or an engine. The AGM-84D is the Harpoon model currently used by the B-52H.(Brian C. Rogers)

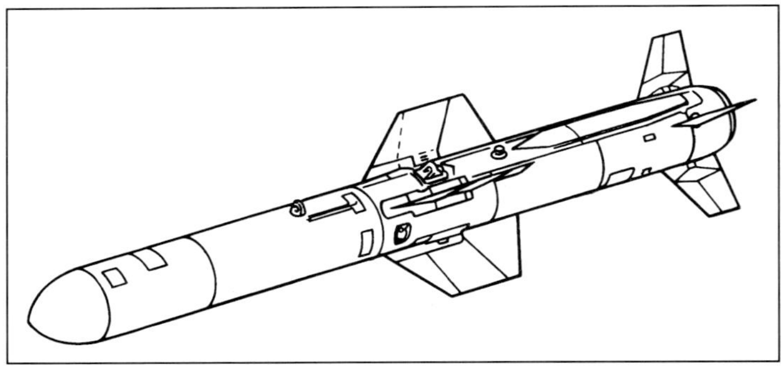

AGM-86B AIR LAUNCHED CRUISE MISSILE (ALCM) - (B-52) and AGM-86C CONVENTIONAL AIR LAUNCHED CRUISE MISSILE (CALCM) - (B-52)

The AGM-86B and C air-launched cruise missiles were developed to increase the effectiveness of B-52 bombers by giving them a long range standoff missile. In February 1974, the Air Force entered into contract with The Boeing Company to develop and flight-test the prototype AGM-86A air-launched cruise missile to be used in the B-1A. This missile was slightly smaller than the later AGM-86B and C models. The AGM-86A model did not go into production. Instead, in January 1977, the Air Force and Boeing began full-scale development of the AGM-86B. Production of the initial 225 AGM-86B missiles began in Fiscal Year 1980 and production of a total 1,715 missiles was completed in October 1986. The air-launched cruise missile had become operational four years earlier, in December 1982, with the 416th Bomb Wing, Griffiss AFB, New York.

B-52H bombers carry six AGM-86B missiles under each wing on each of two externally mounted pylons and eight missiles on a bomb bay rotary launcher. The AGM-86B is powered by a turbofan jet engine. After launch, the folded wings, tail surfaces and engine inlet deploy. At subsonic speeds, the missile then is able to fly complicated routes to a target through use of a terrain contour-matching guidance system. During flight, this system compares surface characteristics with maps of the planned flight route stored in on-board computers to determine the missile's location. As the missile nears its target, comparisons become more specific, guiding the missile to target with pinpoint accuracy.

In June 1986, the first AGM-86C missiles were made by converting AGM-86B to carry a high-explosive blast/fragmentation warhead and replacing the missile's terrain contour-matching guidance system with a Global Positioning System (GPS) navigation system. The C model became operational in January 1991 at the beginning of Operation Desert Storm. B-52Gs, flying "round-robin" missions from Barksdale AFB, launched AGM-86C against high-priority targets in Iraq. These missions marked the beginning of the air campaign for Kuwait's liberation and are the longest known aircraft combat sorties in history (more than 14,000 miles and 35 hours of flight). The B-52H has launched AGM-86Cs in combat. On September 3, 1996, 60-0014 and 60-0059 launched a total of thirteen CALCMs against Iraq. On December 17, 1998, 15 B-52Hs launched approximately 70 CALCMs against targets in Iraq.

The forward three stations visible on this photo of a cruise missile pylon under the left wing of a B-52H are loaded with non-functional AGM-86 load training shapes. (Author)

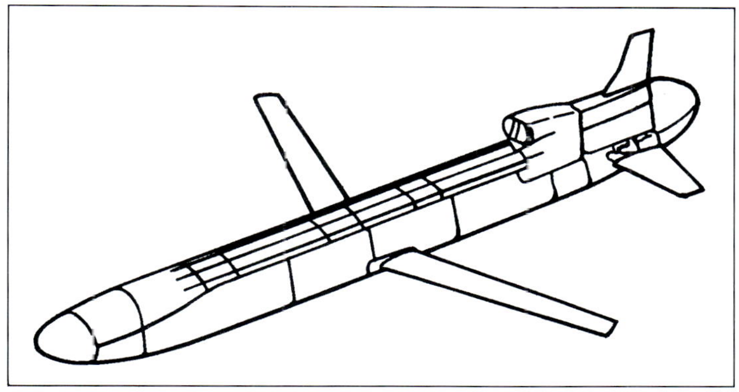

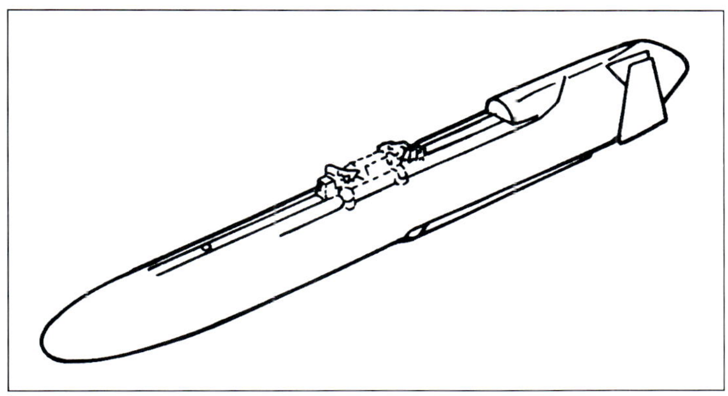

Primary Function: Air-to-surface strategic missile
Engine: Williams Research Corp. F-107-WR-10 turbofan engine
Thrust: 600 pounds (270 kilograms)
Length: 20 feet, 9 inches (6.29 meters)
Weight: 3,150 pounds (1,417.5 kilograms)
Diameter: 24.5 inches (62.23 centimeter)
Wingspan: 12 feet (3.64 meters)
Warhead: AGM-86B – Nuclear AGM-86C – Mod 0 2,000 pound High Explosive, Mod 1 3,000 pound High Explosive
Range: AGM-86B: 1,500-plus miles(1,305 nautical miles); AGM-86C: Classified
Speed: Approximately 550 mph (Mach 0.73)
Guidance System:: AGM-86B Litton inertial navigation element with terrain contour-matching updates AGM-86C Global Positioning System

AGM-129 ADVANCED CRUISE MISSILE (ACM) (B-52H)

The AGM-129 is a subsonic, turbofan-engine powered, winged missile. The ACM can deliver a nuclear warhead in an air-to-ground mission with a very high degree of accuracy at long range. During captive flight, the missile's flight surfaces (wings and fins) are folded or retracted in a stowed position. After launch, the missile's flight surfaces are deployed, and the engine provides thrust within a few seconds. Computer controlled navigation directs the missile to its target. A W80-1 nuclear warhead is installed in the missile weapon bay.

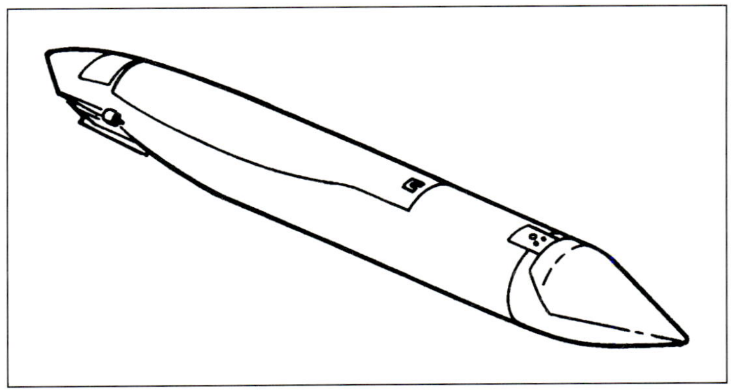

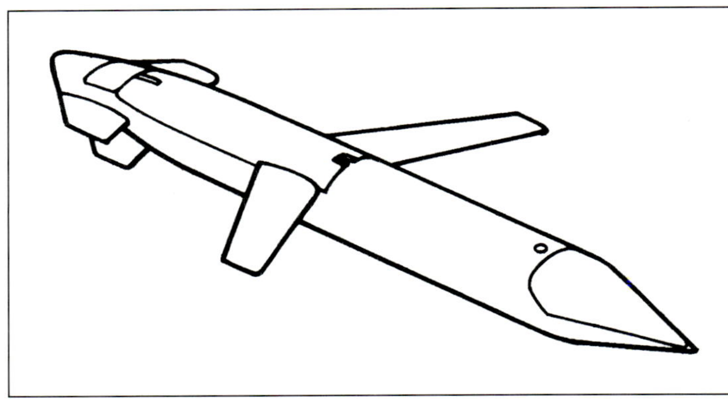

MISSILE DIMENSIONS

Fuselage length	20 feet 10 inches
Fuselage width	2 feet 5 inches
Fuselage depth	2 feet 1.3 inches
Wing Span	10 feet 2.8 inches
Missile Gross Weight	3600 pounds

In this photo, a self propelled munitions lift trailer carrying a cruise missile pylon loaded with AGM-129 ACMs is being positioned under the right wing of a B-52H. (Don Abrahamson)

AGM-142 HAVE NAP MISSILE (B-52)

The AGM-142 (Have Nap) missile is a stand-off, air-to-ground, electro-optical guided weapon. The precision guided, 3000 pound missile is propelled by a single-stage, solid propellant rocket motor. It is armed with a conventional, general purpose or a penetration warhead, and uses either an Imaging Infrared (IIR) or Television (TV) seeker. The AN/ASW-55 Data Link Pod (DLP) provides control to the AGM-142. The DLP receives video and command data, and it transmits commands from the aircraft to the missile.

Of ACC's three bomber types, only the B-52 carries the AGM-142 missile. It is carried externally on the Heavy Stores Adapter Beams (HSAB) under each wing. Each HSAB can carry two missiles, or one missile and a DLP. The DLP is normally carried on the aft station of the right HSAB. For single ship operations, the Buff carries three missiles and a DLP. For formation deliveries, total of four missiles can be carried by each aircraft with at least one aircraft in the formation carrying the DLP and serving as the missile controller for the whole flight. Like the AGM-84 Harpoon, the AGM-142 was originally carried by the B-52G. Under the Rapid Eight program four aircraft (60-0014, 60-0025, 60-0062, and 61-0004) were modified to carry the AGM-142 until the operational B-52H fleet received its modification.

MISSILE TYPES

Missile	Seeker	Warhead
AGM-142A	TV	General Purpose
AGM-142B	IIR	General Purpose
AGM-142C	TV	Penetration
AGM-142D	IIR	Penetration

60-0062, in this photo, is carrying an AGM-142 on the center forward station of the left wing HSAB. (USAF)

PHYSICAL CHARACTERISTICS

Missile Length:	15 feet 10 inches
Diameter:	21 inches
Upper wing span:	5 feet 2 inches
Lower wingspan:	5 feet 8 inches
Weight:	3000 pound class

The AGM-142 is made up of a seeker section, electronics section, warhead section, rocket motor, control section, aerodynamic control surfaces and data link section.

BELOW: An AGM-142 is seen here just after launch from 61-0010 assigned to the 20th Bomb Squadron, 2nd Bomb Wing at Barksdale.

The seeker section has either a TV camera with a zoom lens or an IIR seeker with wide and narrow fields of view. The electronics section which contains the computer system, the inertial measurement unit, and the missile power supply, provides navigation, video processing, antenna pointing, and guidance. The payload for the operational AGM-142A/B is a general-purpose, blast, fragmentation 750 pound class warhead. The payload for an AGM-142C/D is a 770 pound penetration warhead. The solid propellant motor has a nominal burn time of just over two minutes.

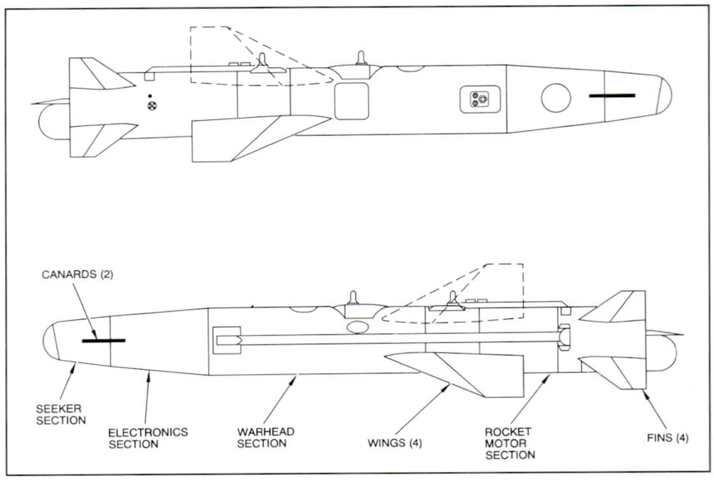

(Paul Hart)

(Brian C. Rogers)

AN/ASW-55 DATA LINK POD (DLP)

Length: 11 feet 7 inches
Diameter: 21 inches
Weight: 865 pounds

The DLP receives video and command data and transmits between the aircraft and the missile during captive and free flight.

Presently the following twelve B-52Hs are modified with the capability to carry, launch and guide the AGM-142 Have Nap:

60-0011 60-0031 60-0062
60-0014 60-0043 61-0002
60-0020 60-0050 61-0004
60-0025 60-0057 61-0010

(Paul Hart)

AGM-154 JOINT STAND OFF WEAPON (JSOW) (B-52, B-1, B-2)

The AGM-152 JSOW is a family of air-to-surface glide weapons, in the 1,000-lb class, provid ng standoff capabilities from 15 nautical miles (low altitude launch) to 40 nautical miles (high altitude launch). The JSOW will be used against a variety of land and sea targets and will operate from ranges outside enemy point defenses. The JSOW is a launch and leave weapon that employs a Global Positioning System (GPS)/ Inertial Navigation System (INS), and is capable of day/night and adverse weather operations. The weapon will be deployed from the B-52, the B-1, and the B-2.

There are three variants of the JSOW; AGM-154A, AGM-154B, and AGM-154C. Each uses a common air vehicle with different payloads. The AGM-154A and AGM-154B will be used by USAF aircraft including the bombers. The AGM-154C is being developed for use by the U.S. Navy.

(Author's Collection)

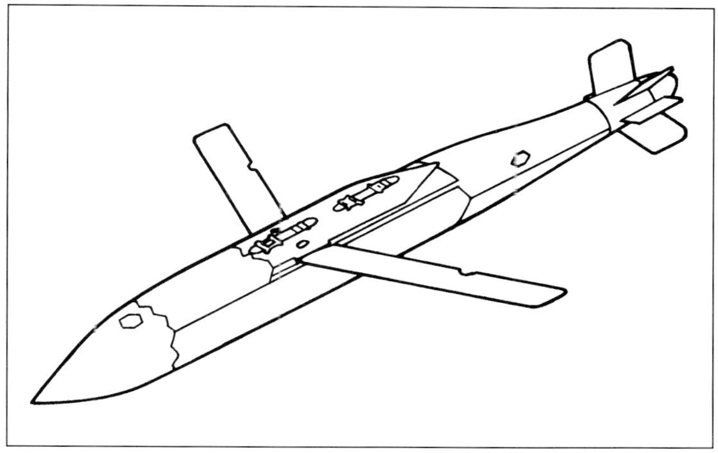

AGM-154A (Baseline JSOW)

The warhead of the AGM-154A consists of 145 BLU-97/B submunitions (same submunition as used in the CBU 87/B). Each bomblet is designed for multi-target in one payload. The bomblets have a shaped charge for armor defeat capability, a fragmenting case for material destruction and a zirconium ring for incendiary effects.

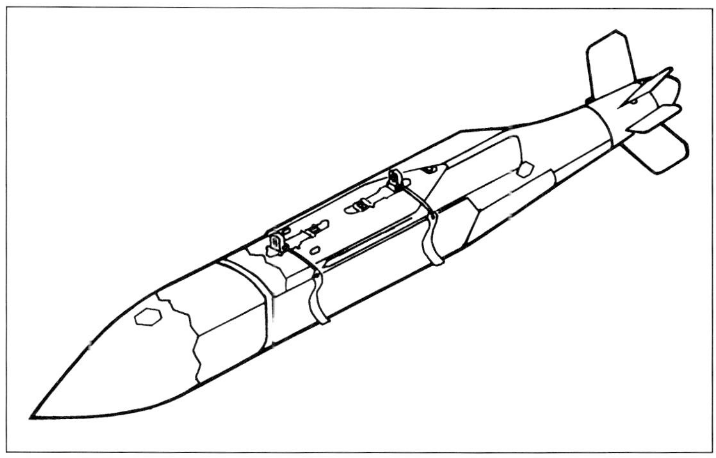

AGM-154B

The warhead for the AGM-154B is the BLU-108/B from the Air Force's Sensor Fuzed Weapon (SFW) program. The JSOW will carry six BLU-108/B submunitions. Each submunition releases four projectiles (a total of 24 per weapons) that use infrared sensors to detect targets. Upon detection, the projectile detonates, creating an explosively formed, shaped charge capable of penetrating reinforced armor targets.

Primary Function: Standoff Outside Point Defense (SOPD) missile
Engine: None
Missile Length: 13 feet 4 inches
Diameter: 22 inches
Weight: 1,000 pound class

AGM-158 JOINT AIR TO SURFACE STANDOFF MISSILE (JASSM) (B-52, B-1, B-2)

The AGM-158 JASSM was a Fiscal year 1996 joint program with Air Force lead and Navy participation. The program replaced the Tri-Service Standoff Attack Missile (TSSAM), canceled for escalating program cost and lack of success. After the termination of the TSSAM, the U.S. Air Force and Navy continued to reiterate the need for a high survivability stand off weapon capable of attacking a variety of deep interdiction type targets.

JASSM is a precision stealthy cruise missile designed for launch from outside area defenses to kill hard, medium-hardened, soft, and area type targets. The B-52, the B-1, and the B-2 are scheduled to receive the weapon. The weapon is capable of attacking both fixed and relocatable targets. The programmed weight is in the 2000 pound class, with a length of fourteen feet (allowing it to fit on the B-1's rotary launcher).

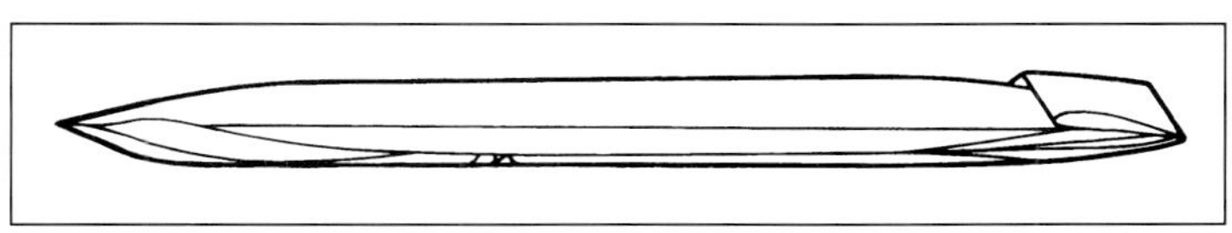

NUCLEAR GRAVITY BOMBS

All three of ACC bombers have the capability to carry and release nuclear gravity bombs. The B-61 and the B-83 are the two modern gravity nuclear bombs carried by ACC's bombers. The two bomb types are carried on rotary launchers internally in the bombers. The B-52 is capable of carrying eight gravity nuclear bombs on the CSRL. The B-1 currently does not have a nuclear mission, but it can carry eight gravity nuclear bombs in each of the three stores bays. The B-2 can carry eight gravity nuclear bombs in each of its two bays.

B-61

The B-61 is the most numerous of the gravity nuclear bombs with an estimated 2,400 units in the U.S. inventory. Development of this light weight, 710 pound, thermonuclear weapon was begun in the mid 1950's. It was originally designed for use by tactical aircraft. The weapon has a yield in the high Kiloton class (variable from 100 to 500 KT). The weapon is capable of release at supersonic speeds and is slowed by a high-speed Kelvar parachute to allow it to "soft land" on the target and is then set off by a timer. A new penetrating version, the B-61 Mod 11, is being developed for the B-52 and the B-2.

B-83

Development of the B-83 was begun in 1974. It is the higher yield of the of the two bombs with an estimated yield in the Megaton class. Six-hundred-fifty B-83's are estimated to be in the U.S. inventory. It is the heavy weight weapon estimated at 2,408 pounds.

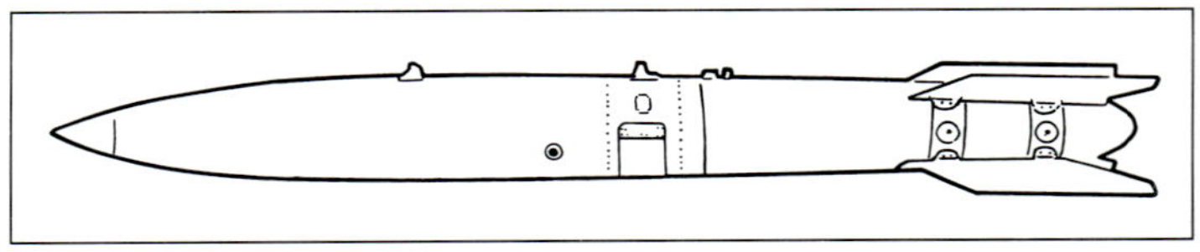

B-61

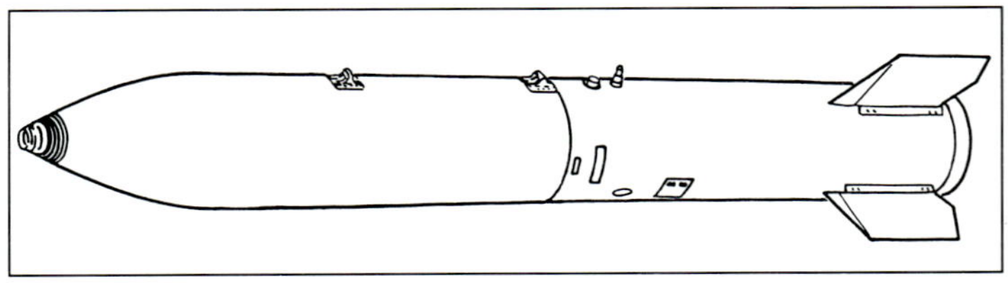

B-83

GUIDED WEAPONS (GBU's)

The new family of guided weapons used on ACC's Bomber Triad uses GPS satellites to aid the weapon in steering to the target. These GPS munitions are the 2000 pound class Joint Direct Attack Munitions (GBU-31 series JDAM). Two additional GBUs, carried only by the B-2, are the 2000 pound GBU-36/B and the 4600 pound GBU-37/B.

GBU-31 JOINT DIRECT ATTACK MUNITION (JDAM) (B-52, B-1, B-2)

The Joint Direct Attack Munition (JDAM) is a joint service weapon which uses a Global Positioning System (GPS) aided Inertial Navigation System guidance set with a MK 84 or BLU-109 gravity delivered bomb. The guidance set, when combined with a warhead, and appropriate fuzes, forms a JDAM Guided Bomb Unit (GBU). The JDAM guidance set gives these bombs adverse weather capability with increased accuracy. The JDAM weapon can be delivered from standoff ranges at high or low altitudes against a variety of land and surface targets during the day or night. When employed, the weapon is released ballistically and is autonomously guided to a designated target using the weapon's on board GPS-aided INS guidance system.

There are seven variants of the weapon, three are USAF versions and four are USN/USMC versions. The USAF bombers use two of the three USAF types; the GBU-31(V)1/B, which consists of a MK-84 general purpose warhead and a JDAM guidance set, and the GBU-31(V)3/B which consists of a BLU-109 penetrator warhead and a JDAM guidance set.

The JDAM guidance sets consist of a tail assembly and a set of mid-body aero-surfaces or "strakes." The tail assembly includes the guidance electronics, the GPS receiver, and control fins. The aero-surfaces or "strakes" consist of three formed panels that attach to the applicable bomb body to provide aerodynamic lift and maneuverability. The BLU-109 variant includes a Hardback, which is a cast aluminum structure designed to provide the appropriate interface with the delivery aircraft.

JDAM provides the capability to deliver multiple weapons onto a single horizontal target aimpoint. The guidance set can control the terminal trajectories of the JDAM weapons to achieve impact angles of 60 degrees and greater.

The JDAM weapon requires targeting data from the releasing aircraft to establish target location and weapon impact parameters. Prior to weapon release, a Target Data Set (TDS) is selected as the target of choice.

60-0050, the B-52H assigned to the 412th Test Wing at Edwards AFB, is on a JDAM test mission. (USAF)

The orange painted parts on this JDAM make up the kit which modifies a MK 84 into a GBU-31(V)1/B. (Author)

This closeup shows the strake aerosurfaces which stabilize the GBU-31(V)1/B as it is steered by the tail fins. (Author)

Six GBU-31(V)1/Bs are seen here under the wing of 60-0050 prior to a flight from Barksdale AFB. (Brian C. Rogers)

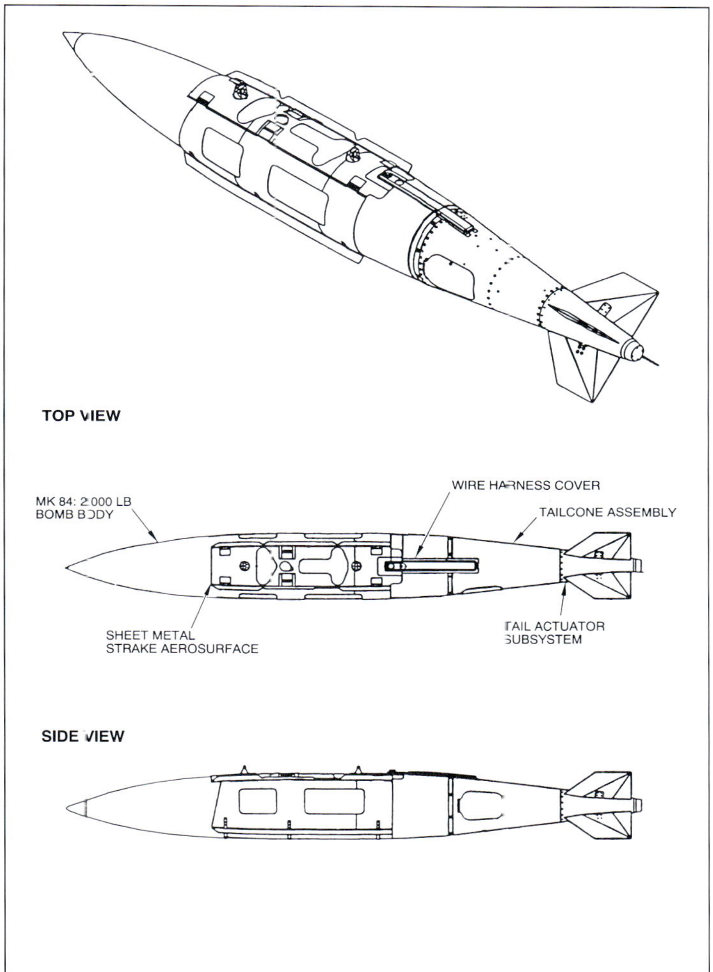

GBU-31(V)1/B

This closeup shows the strake aerosurfaces which stabilize the GBU-31(V)-3/B as it is steered by the tail fins. (Author)

A GBU-31(V)3/B, the JDAM which uses the BLU-109 Penetrator warhead is seen here on this weapons trailer. (Brian C. Rogers)

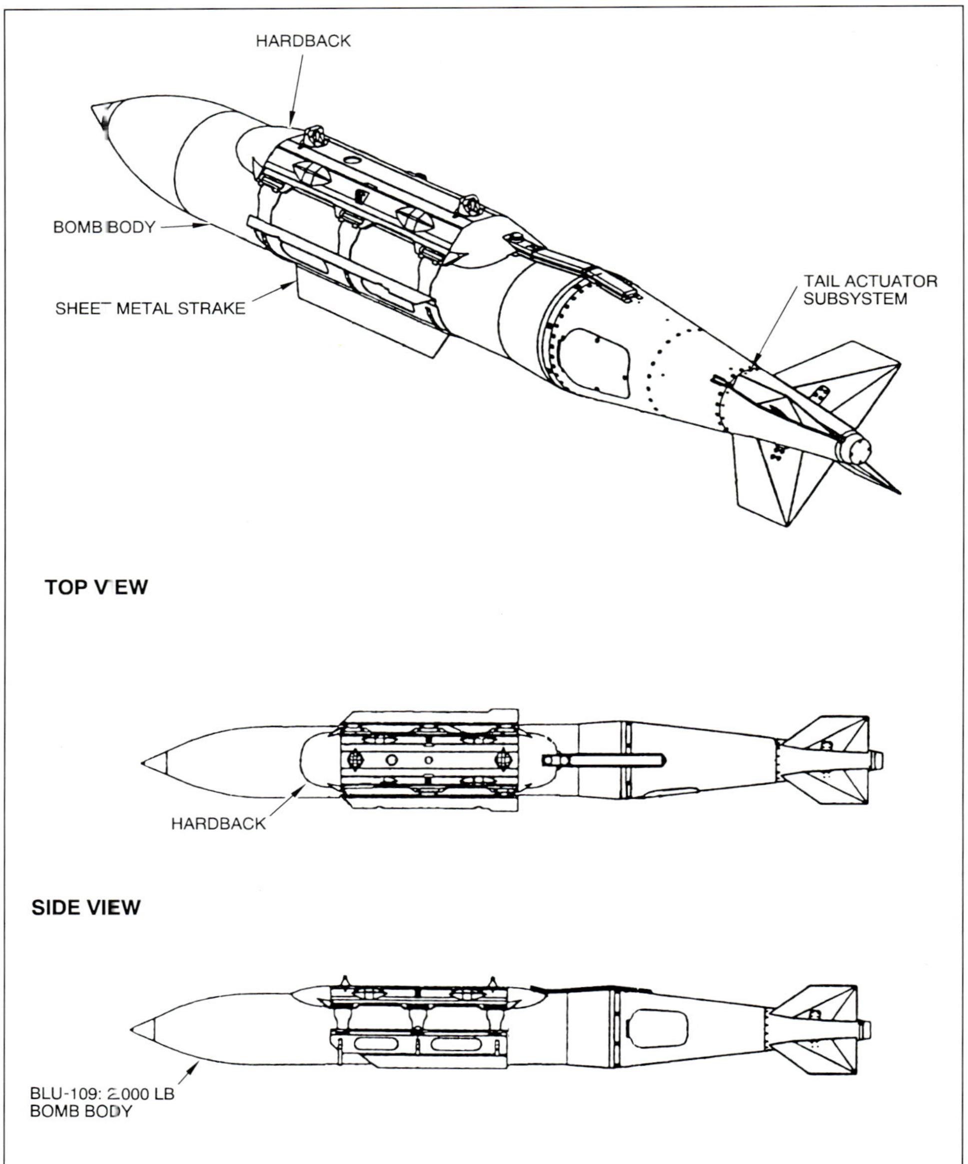

GBU-31(V)3/B

GBU-36/B (B-2)

The GBU-36/B is a MK 84 2000 pound general purpose bomb with a GPS aided munition tail kit (WCU-25/B) and with a boundary layer stabilizing sleeve (MAU-207/B) over the nose. The GPS aided munition tail kit consists of a tail cone containing a guidance and control unit and tail fin control actuator unit. The guidance and control unit has a ring laser gyro INS, a processor, and an embedded GPS receiver. Two GPS antennas receive signals during the bomb trajectory. A weapons system interface connector is secured to the bomb body by a metal strap and connected to the guidance and control unit. The tail control actuator unit consists of actuators, actuator electronics, four tail fins and two thermal batteries. The batteries provide power to the guidance control unit and the tail control actuator unit. The stabilizing sleeve has two aerodynamic boundary layer strips. The boundary layer strips increase the GBU-36/B controllability and stability during high angles of attack. The bomb is fuzed with FMU-139A/B nose and/or tail fuze and the FZU48/B initiator.

This B-2 is releasing a GBU-36/B 2000 pound GPS guided bomb (formerly known as the GATS/GAM). (Northrop-Grumman)

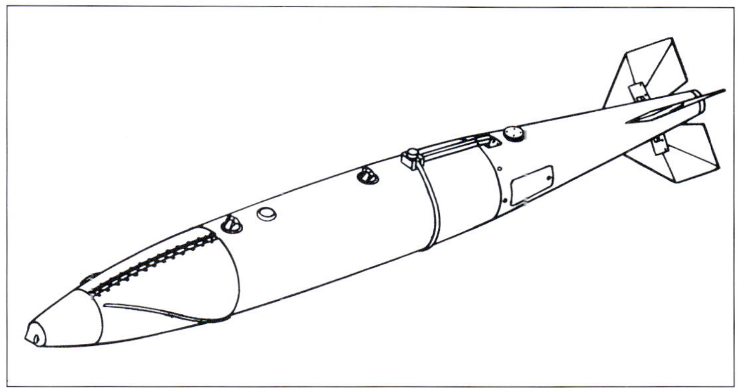

GBU-36/B

The B-2 in this photo is releasing a GBU-37/B 47000 pound GPS guided bomb (formerly known as the GATS/GAM-113). (Northrop-Grumman)

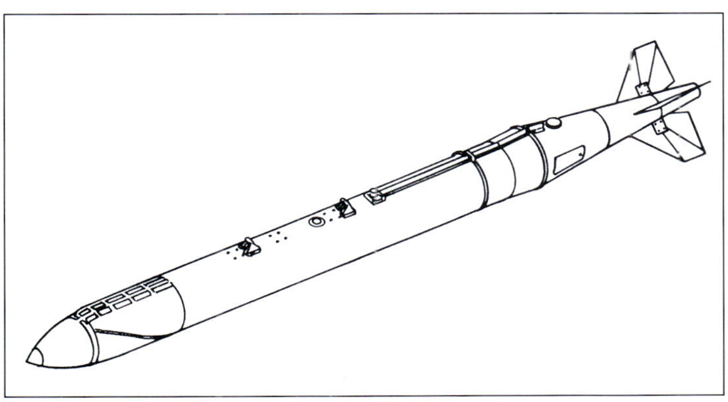

GBU-37/B (GATS/GAM-113)

GBU-37/B (B-2)

The GBU37/B is a 4000 pound penetrator bomb with a GPS aided munition tail kit and stabilizing sleeve. The tail kit is the same as the GBU-36/B with modified software and control fins. The control fins for the GBU37/B are larger for more control authority. The GBU-37/B is fuzed with a FMU-143F/B, G/B, or H/B and a FZU-32B/B initiator.

GBU-36/B and GBU-37/B are not to be carried at the same time on the same B-2. Mission software files contain information for only one type of GPS aided munition. The correct mission software files must be loaded to have correct target ranging, stores weight and aircraft center of gravity computations.

LASER GUIDED BOMBS (LGBs) (B-52)

A laser guided bomb (LGB) is a general purpose or penetrator bomb equipped with electronic and mechanical assemblies that provide laser terminal guidance. LGBs have been released during testing by both the B-52 and B-1; however, presently the only bomber authorized to release LGBs is the B-52.

To make an LGB from a standard bomb, a computer control group (CCG) with guidance canards is attached to the front of the bomb to provide steering commands. A wing assembly is attached to the aft end to provide stability and lift. The LGB is a maneuverable free-fall weapon requiring no electronic interconnect to the aircraft. It has an internal semiactive guidance system that detects laser energy and guides the weapon to a target illuminated by an external laser source. The LGB samples the reflected laser energy from the target and corrects its trajectory as it glides. There are three generations of LGBs: Paveway I, II, and III, but only the I and II are carried on the B-52.

(1) Paveway I. The airfoil group (AFG) for Paveway I has fixed wings configured in either the slow speed (big canard and wings) configuration or the high-speed (small canards and wings) configuration.

(2) Paveway II. It is an improved version of Paveway I with folded wings which deploy at release.

GBU-10/B, A/B (MK 84 PAVEWAY I)

This configuration uses either the MAU-157/B or MAU-157A/B computer and the MXU-G00/B AFG. It weighs 2055 pounds, is 168 inches long, and has a body diameter of 18 inches. For the B-52 application, the wing extenders are not installed and the canards are shortened to the high speed configuration. It is carried externally on the HSAB in the 2-0-2 configuration.

GBU-10C/B, D/B, E/B (MK 84 PAVEWAY II)

This store uses either the MAU-169A/B or MAU169B/B computer and the MXU-65IIB AFG. It weighs 2083 pounds, is 170 inches long, and has a body diameter of 18 inches. It is carried externally on the HSAB in the 2-1-2 configuration.

GBU-10G/B, H/B, J/B (BLU-109 PAVEWAY II)

This store uses the BLU-109/B (1-2000) penetrator warhead, the same computers and AFG as the MK 84 Paveway II. It

B-1B 84-0049 is seen here releasing GBU-10G/B Paveway II LGB during testing.(USAF)

This GBU-10/B Paveway I LGB still has the extended fins and canards attached. (Brian C. Rogers)

weighs 2147 pounds, is 169 inches long, and has a body diameter of 15 inches. It is carried externally on the HSAB in the 2-1-2 configuration.

GBU-12B/B, C/B, D/B (MK 82 PAVEWAY II)

This store uses the MK 82 warhead, the MXU650/B AFG, and the MAU-169/B, A/B, or B/B computer. It weighs 624 pounds, is 131 inches long, and has a body diameter of 10.75 inches. It is carried externally on the HSAB in a 2-1-2 weapons configuration.

CLUSTER BOMBS

A Cluster Bomb Unit (CBU) is a completely assembled weapon with integral bomblets (Bomb Live Units [BLU]) loaded into a dispenser (SUU); the combination is designated a CBU. The combination of bomblets and dispenser determine the type of CBU. The dispenser opens after a preset time or at a preset altitude, dispersing the bomblets.

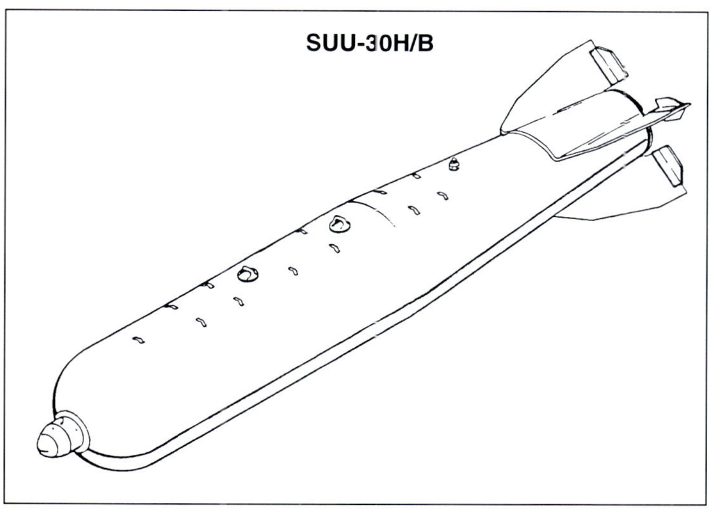

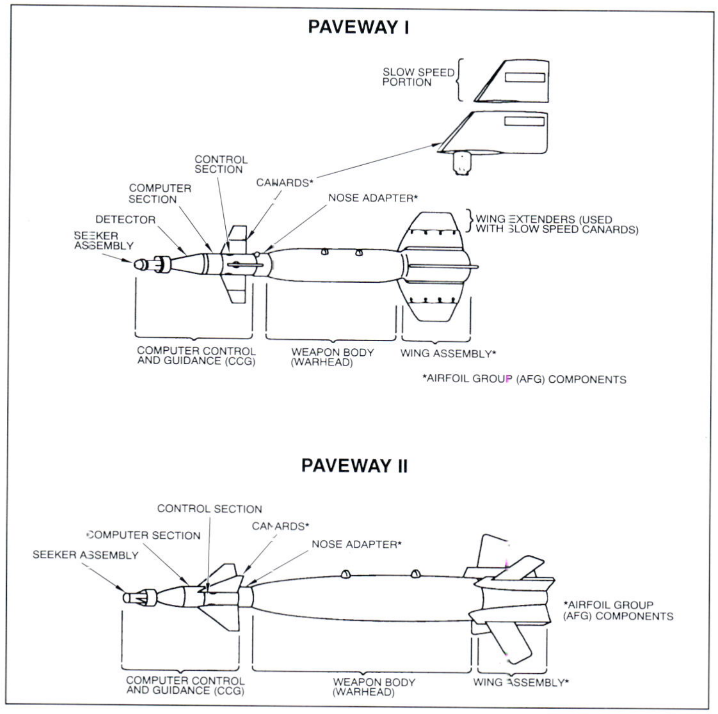

SUU-30H/B DISPENSER WEAPONS (B-52)

The B-52 was the only bomber authorized to release the SUU-30H/B CBUs. As part of the worldwide prohibition against land mines, the SUU-30 and its associated CBUs have been removed from the United States inventory. The SUU-30 dispenser is 90 inches long, 16 inches in diameter, features four fixed aluminum fins on the tail section, and has 14 inch lug spacing. A single fuze well is located in the nose. When the fuze functions, the front end of the dispenser blows open. Ram air completes case opening and the case splits into two pieces. The CBU-52B/B, CBU-58/B, and CBU-71/B are SUU-30H/B dispenser weapons.

CBU-52B/B

The CBU-52B/B uses the SUU-30H/B dispenser, weighs 785 pounds, and is loaded with 220 BLU-61A/B. The BLU-61A/B is 3.5 inches in diameter, weighs 2.7 pounds, and explodes on impact.

Nine CBU-58/Bs are seen loaded on this stub pylon on HSAB combination. (Brian C. Rogers)

CBU-58/B

The CBU-58/B uses the SUU-30H/B dispenser, weighs 820 pounds, and is loaded with 650 BLU-63/B. The BLU-63/B is 2.94 inches in diameter, weighs .93 pound, and explodes on impact.

CBU-71/B

The CBU-71/B uses the SUU-30H/B dispenser, weighs 820 pounds, and is loaded with 650 BLU-86/B. The BLU-86/B is 2.94 inches in diameter, weighs .93 pound, and has an explosive charge of.25 pound. Random time delay fuzing is used.

SUU-64/B AND SUU-65/B TACTICAL MUNITIONS DISPENSERS

The Tactical Munitions Dispenser (TMD) is a cluster munitions dispenser. The TMD is made in two major versions: SUU-64/B and SUU-65/B. The major difference between the two versions is in the tail sections. The SUU-64/8 has a nonspin tail section, and the SUU-65/B has a fin cant mechanism which provides aerodynamic spinning of the dispenser and enhances submunition dispersion.

SUU-64/B

The SUU-64/B is made up of three main assemblies: body, tail, and nose. The nose includes an integral timer fuze and an optional FZU-39/B sensor. The body is an aluminum cylinder that is welded to the forward bulkhead. Three longitudinal strands of Aluminum Linear Shaped Charges (ALSC), and a circumferential strand of ALSC at the aft bulkhead function to cut the dispenser body into three longitudinal pieces and separate the tail section. The aft bulkhead provides the primary structural support for the aft end of the body and acts as a seal for the cargo section.

TMD function mode is determined by the arming solenoid selection made by the releasing aircraft. The nose solenoid is used to pull the fuze arming lanyard and the tail solenoid is used to pull the fuze option lanyard. When nose/tail arming is selected the timer acts as a safe separation timer and the TMD will not function until the FZU-39/B proximity sensor senses the preset function altitude.

The tail section is attached to the aft bulkhead and has four extendible fins which are held in the retracted position by a fin release band. When the fin release lanyard is pulled, the fins are extended. The fins are then locked in the extended position.

SUU-65/B

The SUU-65/B is a SUU-64/B with a modified tail section that imparts spin to the dispenser. The spin tail differs from the nonspin tail in that it contains a fin cant mechanism and an explosive bolt and tail harness. The spin mode is selected before release. After release, at a selected altitude, a spring-loaded fin cant mechanism rotates the fins to the fully canted position causing the dispenser to rotate. The TMD opens when the preset spin rate is achieved.

This TMD is a dummy unit used as a loading trainer. The extendible fins here are seen stowed. (Author)

CBU-87/B CLUSTER BOMB COMBINED EFFECTS MUNITIONS (B-52, B-1, B-2)

The CBU-87/B Combined Effects Munition (CEM) consists of a SUU-65/B TMD, 202 BLU-97/B bomblets and an optional FZU-39/B proximity sensor. The BLU-97/B case is made of scored steel designed to fragment into approximately 300 performed 30-grain fragments for defeating light armor and personnel. It contains a forward-firing, shaped-charge liner for defeating armor, and a zirconium ring for incendiary capability. An Air Inflatable Decelerator (AID) that provides drag, orientation and flight stability for the bomblet is held encased by a cap called a spyder.

When the BLU-97/B is released into the airstream from the SUU-65 dispenser and attains its minimum airspeed, the airflow releases the spyder which pulls the cup assembly rearward, exposing the AID to the airstream. As the AID is inflated by ram air, it orients and stabilizes the BLU-97/B for proper target impact by despinning the submunition and reducing the descent rate. When the BLU-97/B impacts the target, the submuniton is detonated.

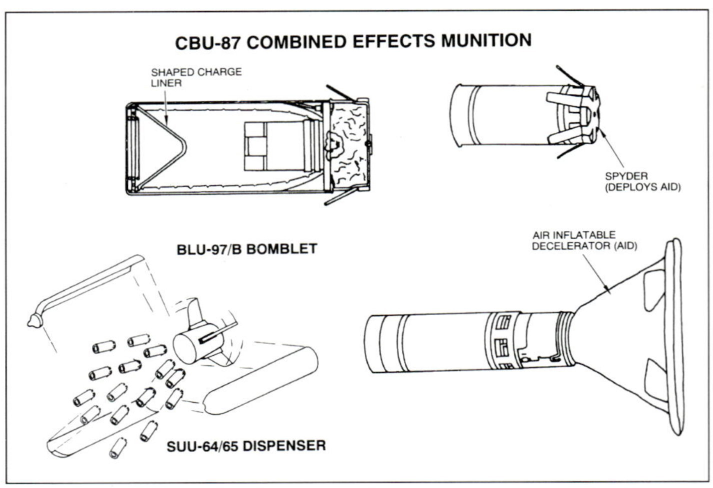

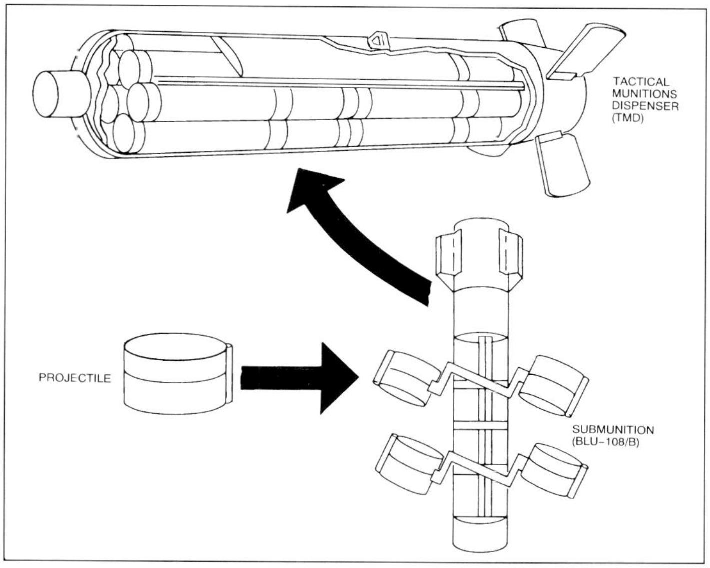

CBU-89 CLUSTER BOMB (GATOR) (B-52, B-1, B-2)

The CBU-89/B GATOR consists of a SUU-64/B TMD, an electronics assembly, 72 BLU-91/B Anti-Tank mines (AT), 22 BLU-92/B Anti-Personnel mines (AP) and an optional FZU-39/R proximity sensor.

Mine arming begins when the TMD opens and requires two minutes to complete. After the mine is armed, mine detonation is initiated by target detection, mine disturbance, low battery voltage and self-destruct time-out. The submunitions are programmed with classified self-destruct settings which are ground selectable on the self-destruct switch located on the TMD.

The AT mine is a magnetic sensing submunition and is effective against tanks and armored vehicles. Detonation occurs when the magnetic sensor detects a target or if the mine is disturbed. If no target is detected prior to expiration of the preset self-destruct time, the mine will detonate. If mine battery voltage drops below acceptable level, the mine will detonate.

The BLU-92/B AP mine has a fragmenting case warhead triggered by trip wires. The trip wires deploy at mine arming. Mine arming is otherwise similar to the BLU-91/B arming process. Detonation occurs when a target actuates the trip wires or when the mine is disturbed. If no target is detected prior to expiration of the preset self-destruct time, the mine will detonate. If mine battery voltage drops below acceptable level, the mine will detonate.

CBU-97/B SENSOR FUZED WEAPON (B-1, B-2)

The CBU-97 Sensor Fuzed Weapon (SFW) consists of a SUU-64/B TMD, an air bag dispensing system, 10 BLU-108/B submunitions packaged in two bays of five, an integral timer fuze, and the FZU-39/B proximity sensor. SFW is designed to provide multiple kill per pass capability against tanks, armored vehicles, artillery, armored personnel carriers, and support vehicles.

The BLU-108/B submunition consists of a nose electronics section, warhead launcher assembly, rocket motor, and Orientation/Stabilization Device (OSD). The launcher assembly holds four submunitions called skeet (four skeet for each BLU-108/B or a total of 40 skeet per SFW munition). The OSD consists of externally mounted fins, a drogue parachute, and a main parachute.

SFW is delivered as an unguided, gravity drop, direct attack or limited standoff munition. After release, the SUU-64/B TMD falls and opens as discussed earlier. When the TMD skin is released, two gas generators fire inside the TMD inflating forward and aft bay air bags. As gas pressure builds inside the air bags they expand ejecting first the five submunitions in the forward bay, followed by the five submunitions in the aft bay. After ejection from the TMD, the submunitions descend to their functioning altitude. At that altitude the SFW rocket motor fires which provides reverse (upward) thrust and spin to the submunition. This upward spin provides centrifugal force to throw the skeet projectiles into a randomly oriented pattern.

The skeet are anti-armor warhead projectiles and function as the kill mechanism of the SFW. Each contains a thermal battery, a passive IR sensor, processing electronics, a safe-arm device, and an explosively formed projectile warhead.

The skeet are canted slightly during articulation so that upon release from the spinning submunition their flight path will contain a slight wobble. This is referred to as coning. As the skeet fly away from the submunition, they are spinning and coning. The spinning arms the skeet via rotation of the safe-arm device while the coning action causes an IR seeker boresighted to the spin axis of the skeet to describe a scan pattern across the ground.

The skeet seeks targets based on predetermined IR signatures. When a target is detected, a processor in the skeet issues a fire pulse to the detonator. The explosive output of the detonator initiates the high explosive which in turn produces a high velocity called an Explosively Forged Penetrator. This penetrates the engine area of the target vehicle from the top of the target inflicting a mobility kill. If the skeet does not detect a target during the flyout period, a self-destruct feature detonates the warhead.

WIND CORRECTED MUNITIONS DISPENSER (WCMD) (B-52, B-1 and B-2 Future Weapons)

The Wind Corrected Munitions Dispenser (WCMD) is a Cluster Bomb Unit (CBU) made up of a Tactical Munitions Dispenser (TMD) with a WCMD tail assembly and a FZU-39/B proximity sensor. The WCMD tail assembly guides the dispenser to a preprogrammed functioning point above the target using an Inertial Measuring Unit (IMU).

CBU-103 (CEM)

The CBU-103 Combined Effects Munition (CEM) series cluster bomb consists of a SUU-65 bomb dispenser, 202 BLU-97 series fragmentation bombs, and a WCMD guided tail kit. This variant is guided from release to a function point above the target as determined by the fuze settings. At the preset height above the target, the FMU-39 sends a signal to the tail, commanding the fins to spin the dispenser. Once the dispenser reaches a preset RPM (determined by another fuze setting), the fuze initiates a linear cutting charge which opens the TMD and disperses the submunitions.

CBU-104 (GATOR)

The CBU-104 Gator series cluster bomb consists of a SUU-64/B bomb dispenser, 94 submunitions (72 BLU-91/B antitank mines and 22 BLU-92/B antipersonnel mines), and a WCMD guided tail kit. This variant is guided from release to a function point above the target as determined by the fuze settings. At the preset height above the target, the FMU-39 sends a signal to the fuze, initiating a linear cutting charge which opens the TMD and disperses the submunitions.

CBU-105 (SFW)

The CBU-105 Sensor Fuzed Weapon (SFW) series cluster bomb consists of a SUU-66/B bomb dispenser with integral fuze, a FZU-39/B, 10 BLU-108 series submunitions, and a WCMD guided tail kit. This variant is guided from release to a function point above the target as determined by the fuze settings. At the preset height above the target, the FMU-39 sends a signal to the fuze, initiating a linear cutting charge the TMD. The BLU-108 release and deploy to cover an oval pattern on the ground.

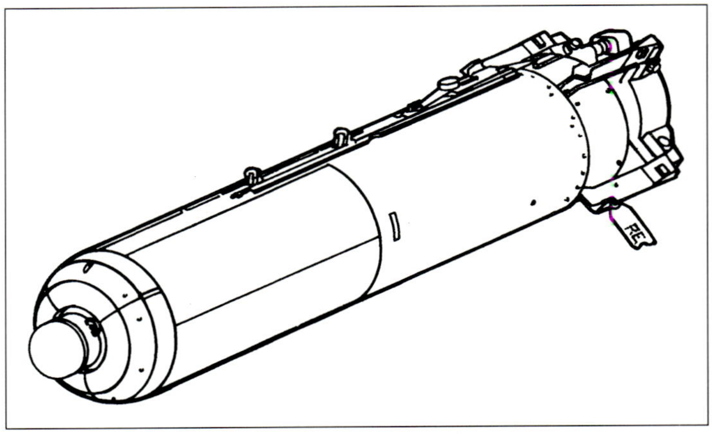

Six test WCMDs are seen here under the wing of 60-0050 prior to a test flight. (Author's Collection)

This is a closeup of the WCMD tail assembly which guides the weapon to the target area. (Author's Collection)

GENERAL PURPOSE BOMBS

MK 84 LDGP (B-52, B-1, B-2)

The 2000 pound low-drag general purpose (LDGP) MK 84 LDGP bomb has a cylindrical body with a conical tail fin. The bomb is filled with 945 pounds of Tritonal or H-6 high explosive.

MK 82 LDGP (B-52, B-1, B-2)

The 500 pound low-drag general purpose MK 82 LDGP bomb has a thick case, streamlined cylindrical body with a conical tail fin. The bomb is filled with 192 pounds of Tritonal high explosive.

MK 82 AIR (B-52, B-1, B-2)

The 500 pound MK 82 AIR (air inflatable retarder) bomb is used for high-speed, low-altitude release. The MK 82 AIR is a MK 82 bomb body with a combination balloon and parachute retarder (ballute) attached. The tail consists of a finned stabilizer canister. The canister encloses the ballute for aerodynamic stability.

M117 GP (B-52, B-2)

The 750-pound general purpose M117 GP (low-drag) bomb is used against a variety of targets. The M117 GP has an ogival shaped nose, a cylindrical body, a tapered aft section with conical fin and is filled with 386 pounds of Tritonal or Minol II high explosive.

Five MK 84 General Purpose 2000 pound bombs are seen here loaded on the stub pylon HSAB combination in the 2-1-2 configuration. (Brian C. Rogers)

M117/M117GP/M117LD (B-52, B-2)

The M117/M117GP/M117LD is fitted with the MAU-103 low drag conical fin. The MAU-103 differs from a standard conical fin in that its stabilizer blades extend almost the entire length of the tail cone.

This display at Barksdale was laid out to show the weapon types capable of being carried by the B-52H. (USAF)

M117AIR (B-52)
The M117AIR is the designation of the M117GP when fitted with the BSU-93 air inflated retarder (AIR) fin.

M117R (B-52, B-2)
The M117R (retarded) bomb has an M117 bomb body with a metal retarding tail fin. The bomb is used in a high-speed, low-altitude release.

Nine M117AIR 750 pound bombs are seen here loaded on a stub pylon HSAB combination. (Brian C. Rogers)

BELOW: The B-2 is seen here dropping M117R retarded 750 pound bombs. (USAF)

SEA AND LAND MINES

The aircraft have the capability to deliver a wide variety of mines. Some mines are strictly anti-ship or anti-submarine devices designed for employment in varying depths of water. Others are for use only in shallow water or released on land for area denial. Sea mines are typically designed for one of two types of employment. Bottom mines are designed for relatively shallow water and lie on the bottom of the sea, harbors, or inland waterways. Moored mines are designed for deep water employment. Moored mines have an anchor, deployment cable, and depth sensors. When the mine is laid, it sinks to the bottom, releases its anchor, and lets out the deployment cable until its depth sensor detects the preset deployment depth.

Mines use a number of different sensors (magnetic, hydrostatic, acoustic, etc.) to detect targets and detonate when specified parameters are met. All mines are battery operated and feature either self-destruct or sterilization features which function prior to end of battery life. All mines have some form of retarding device to lessen the impact shock of delivery. The retarding device for moored mines is typically a parachute pack and a set of fins which mount on the aft end of the mine. Shallow water bottom mines or land mines typically have either a parachute retarder, a metal retarding fin like a MK 82 Snakeye bomb, or a ballute retarder. The flight gear (fins) for internally carried moored mines is slightly different than that for externally carried moored mines. Externally carried mines may have an aerodynamic nose fairing installed on the front of the mine. When the mine impacts the water, the nose fairing breaks free of the mine.

Different color schemes identify mines. Live US Navy mines are painted a dark brick red. Training mines, suitable for carriage and release, are painted with wide orange and white stripes. Load trainers, not suitable for flight, are painted gold. Quickstrike mines are typically olive drab with distinctive colored striping to distinguish them as general purpose bomb bodies configured as mines.

MK 52 mines are seen loaded in bomb bay of this B-52H (Brian C. Rogers)

A weapons is seen steadying a MK 52 mine during loading (Brian C. Rogers)

BOTTOM MINES (B-52)

The MK 52 and MK 55 mines are no longer in the service inventory. Inert versions are used for exercise training only.

MK 52

The MK 52 was designed to detect and destroy submarines or surface vessels. The MK 52 is 70 inches long, 18 inches in diameter, and weighs 1266 pounds, depending upon the type of modification. Its HBX-1 explosive weighs 625 pounds.

MK 55

The MK 55 was designed to detect and destroy submarines or surface vessels. The MK 55 is 89 inches long, 23 inches in diameter, and weighs 2200 pounds.

MOORED MINES (B-52)

MK 56

The MK 56 is designed specifically to counter high speed, deep operating submarines. The MK 56 is 114.5 inches long, 22.35 inches in diameter, and weighs 2150 pounds without tail fin and nose fairing and 2215 pounds with tail fin and nose fairing. Its HBX-5 explosive weighs 360 pounds and uses a magnetic firing mechanism featuring a total field (three dimensional) detector.

Flight gear for carriage and release of the MK 56 mine is a MK 28 parachute pack with MK 112 actuator and drag plate. When the mine is released from the aircraft, an arming wire is withdrawn, and the firing pin fires the ballistic charge, initiating parachute deployment.

MK 60 Captor

The MK 60 is an anti-submarine mine designed for mooring in deep water. It is 145 inches long, 21 inches in diameter, weighs 2366 pounds, and features an influence target detection system. When a target is detected, the device releases a modified MK 46 torpedo that swims out of its capsule and attacks the target. Flight gear for carriage and release of the MK 60 consists of a MK 42 parachute pack with a MK 198 actuator. When the mine is released from the aircraft, an arming wire is withdrawn, and the firing pin fires the ballistic charge, initiating parachute deployment. Four fixed fins extend from the parachute housing for stability after release.

QUICKSTRIKE (QS) BOTTOM MINES

The QS class of mines can be effectively employed in shallow water. The QS mine is a standard general purpose bomb fitted with a modular kit of components (QS kit). The kit has a mine arming, detection, and firing system. A hydrostatic sensor (part of the arming mechanism) must detect that the mine is submerged or the weapon will not arm. Although the same component kit is used for all QS mines, a variety of bomb bodies and fins constitute QS mine differences.

DST (M117) MK 59, Quickstrike (B-52)

The MK 59 QS uses a M117 bomb body, metal MAU-91 retarding fin, and the QS kit. A butt plate is installed on the rear of the bomb body to make it watertight. When carried internally, a Kevlar stow pack is used as a fin deployment extender for the MK 59.

MK 52 MINE

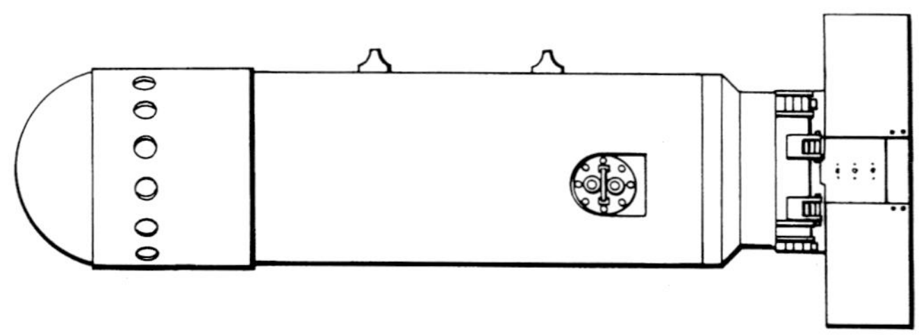

MK 55 MINE

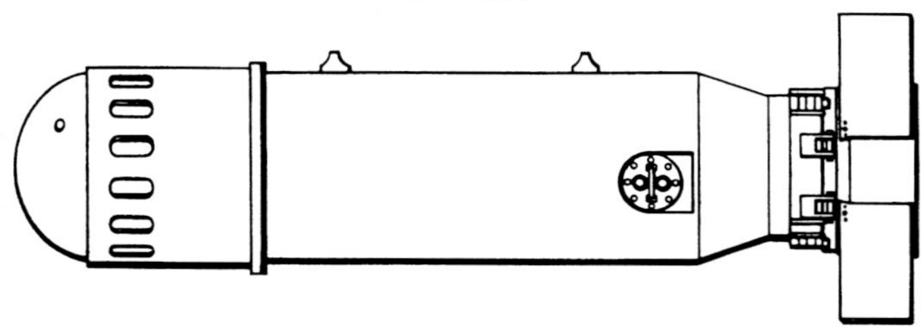

MK 56 MINE

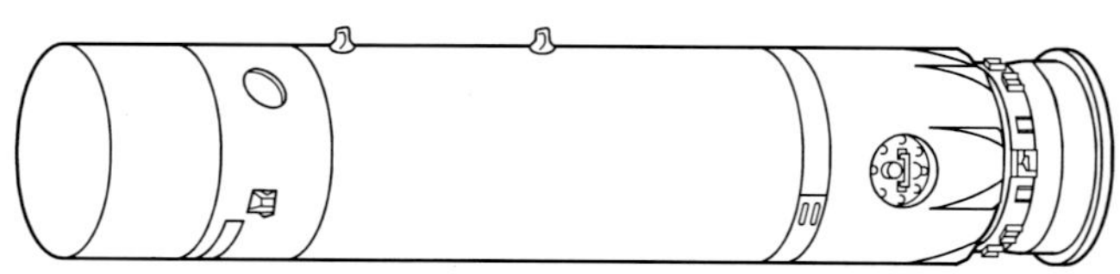

MK 60 CAPTOR MINE

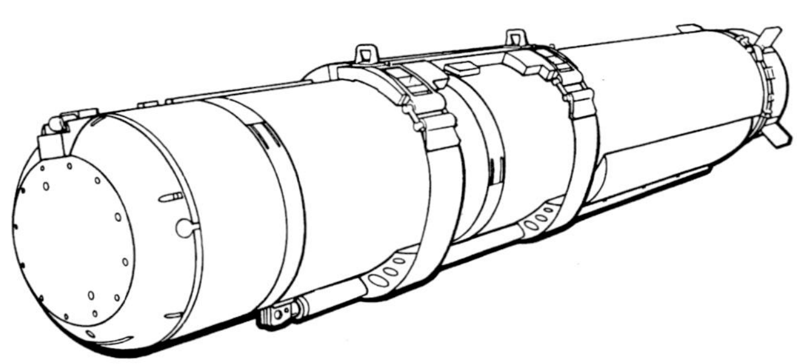

MS-62 QS, Quickstrike (B-52, B-1, B-2)
The MK 62 QS mine is a shallow water bottom mine and uses a MK 82 bomb body, MK-16 tail kit and MK 130 arming kit. The MK-16 tail permits high speed low altitude delivery. The MK-16 uses a parachute retarder. When the parachute deploys, after release, the arming wire is pulled allowing the mine to arm.

The MK 62 QS uses a MK 82 bomb body, metal MK 15 retarding fin, and QS kit. When carried internally, a Kevlar stow pack is used as a fin deployment extender.

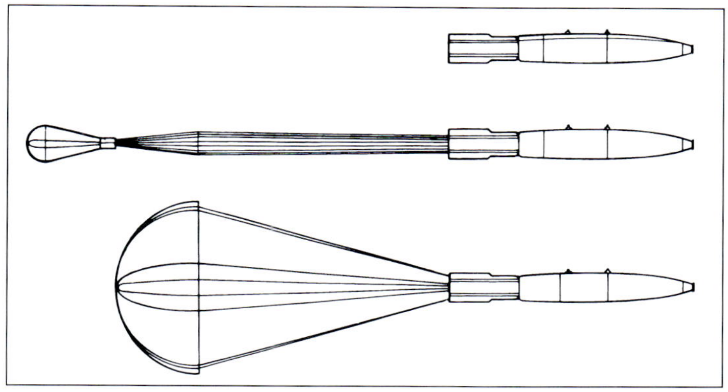

The descent of the MK 62 Quick Strike mine is slowed by the deployment of a drag parachute as seen here. The parachute slows the mine enough to let it safely enter the water and then arm.

MK 63 QS, Quickstrike (B-52)
The MK 63 QS uses a MK 83 bomb body, either the metal MAU-91 retarding fin or the MK 12 para-tail retarder, and the QS kit.

MK 65 QS, Quickstrike (B-52)
The MK 65 QS Mod 0 and Mod 1 mines consist of a new thermally coated, thin-walled body specifically designed as a mine.

This B-1 is releasing MK 62 Quick Strike mines. The MK 62 QS uses the MK 82 500 pound bomb as its warhead. Its fall is retarded by parachute. (USAF)

Also from the Publisher

NORTHROP'S
YF-17
COBRA
A PICTORIAL HISTORY
Don Logan
A SCHIFFER MILITARY HISTORY BOOK

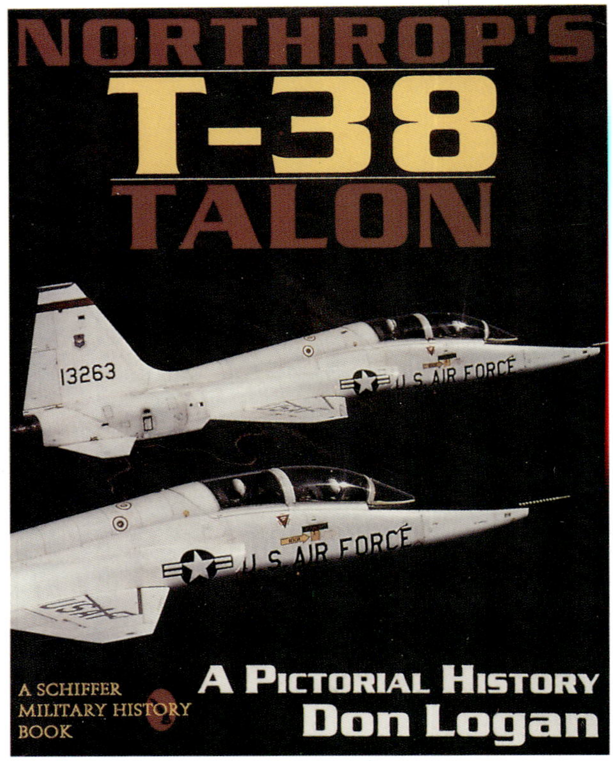
NORTHROP'S
T-38
TALON
A PICTORIAL HISTORY
Don Logan
A SCHIFFER MILITARY HISTORY BOOK

ROCKWELL
B-1B
SAC'S LAST BOMBER
DON LOGAN

REPUBLIC'S A-10
THUNDERBOLT II
A PICTORIAL HISTORY • DON LOGAN